ID678902

KNOTS

ALEXEI SOSSINSKY

TRANSLATED BY GISELLE WEISS

KNOTS

MATHEMATICS WITH A TWIST

HARVARD UNIVERSITY PRESS
CAMBRIDGE, MASSACHUSETTS
LONDON, ENGLAND
2002

Printed in the United States of America

Originally published as *Nœuds: Genèse d'une Théorie Mathématique*

Illustrations by Margaret C. Nelson
Design by Marianne Perlak

Library of Congress Cataloging-in-Publication Data

Sossinsky, A.B.

[Nœuds. English]

Knots : mathematics with a twist / Alexei Sossinsky ; translated by Giselle Weiss.

p. cm.

Includes bibliographical references.

ISBN 0-674-00944-4 (alk. paper)

1. Knot theory. 2. Low-dimensional topology. I. Title.

QA612.2.S6713 2002

514′.224—dc21 2002027295

CONTENTS

PREFACE

Butterfly knot, clover hitch knot, Gordian knot, hangman's knot, vipers' tangle—knots are familiar objects, symbols of complexity, occasionally metaphors for evil. For reasons I do not entirely understand, they were long ignored by mathematicians. A tentative effort by Alexandre-Théophile Vandermonde at the end of the eighteenth century was short-lived,[1] and a preliminary study by the young Karl Friedrich Gauss was no more successful. Only in the twentieth century did mathematicians apply themselves seriously to the study of knots. But until the mid-1980s, knot theory was regarded as just one of the branches of topology: important, of course, but not very interesting to anyone outside a small circle of specialists (particularly Germans and Americans).

Today, all that has changed. Knots—or more accurately, mathematical theories of knots—concern biologists, chemists, and physicists. Knots are trendy. The French "nouveaux philosophes" (not so new anymore) and postmodernists even talk about knots on television, with their typical nerve and incompetence. The expressions "quantum group" and "knot polynomial" are used indiscriminately by people with little scientific expertise. Why the interest? Is it a passing fancy or the provocative beginning of a theory as important as relativity or quantum physics?

This book addresses this question, at least to some extent, but its aim is certainly not to provide peremptory answers to global inquiries. Rather, it presents specific information about a subject that is difficult to grasp and that, moreover, crops up in many guises, often imbued with mystery and sometimes with striking and unexpected beauty.

This book is intended for three groups of readers: those with a solid scientific background, young people who like mathematics, and others, more numerous, who feel they have no aptitude for math as a result of their experience in school but whose natural curiosity remains intact. This last group of readers suffers from memories of daunting and useless "algebraic expressions," tautological arguments concerning abstractions of dubious interest, and lifeless definitions of geometric entities. But mathematics was a vibrant field of inquiry before lackluster teaching reduced it to pseudoscientific namby-pamby. And the story of its development, with its sudden brainstorms, dazzling advances, and dramatic failures, is as emotionally rich as the history of painting or poetry.

The hitch is that understanding this history, when it is not reduced to simple anecdotes, usually calls for mathematical sophistication. But it so happens that the mathematical theory of knots—the subject of this book—is an exception to the rule. It doesn't necessarily take a graduate of an elite math department to understand it. More specifically, the reader will see that the only mathematics in this book are simple calculations with polynomials and transformations of little diagrams like these:

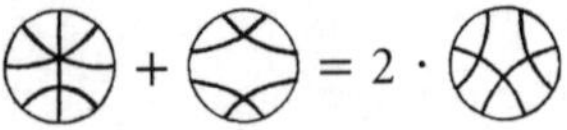

+ = 2 · and − =

Readers will also have to draw on their intuition of space or, failing that, fiddle with strings and make actual knots.

My desire to avoid overly abstract and technically difficult mathematics led me to leave out completely the most classical tool of the theory of knots (and the most efficient at the early stages), the so-called fundamental group. The first successes with the theory—those of the mathematicians of the German school (N. G. Van Kampen, H. Seifert, M. Dehn), the Dane J. Nielsen, and the American J. W. H. Alexander—were based on the judicious use of this tool. Their work will barely be mentioned here.

Given the diversity of the topics tackled in this book, I have not tried to provide a systematic and unified exposition of the theory of knots; on the contrary, various topics are scattered throughout the chapters, which are almost entirely independent of each other. For each topic, the starting point will be an original idea, as a rule simple, profound, and unexpected, the work of a particular researcher. We will then follow the path of his thinking and that of his followers, in an attempt to understand the major implications of the topic for contemporary science, without going into technical details. Accordingly, the chapters are ordered more or less chronologically. But I have striven to minimize cross-references (even if it means repeating certain passages), so that the chapters can be read in whatever order the reader chooses.

Before I review the topics taken up in each chapter, it is worth mentioning that prior to becoming the object of a theory, knots were associated with a variety of useful activities. Of course, those activities are not the subject of this book, but talking a little about their practical charms will make it easier to glimpse the beauty of the theory.

Since Antiquity, the development of knot making was motivated by practical needs, especially those of sailors and builders. For each specific task, sailors invented an appropriate knot, and the best knots survived, passing from generation to generation (see Adams, 1994). To tie a rope to a rigid pole (a mooring or a mast), one uses the clover hitch knot (see Figure P.1a), the rolling hitch knot (b), or the camel knot (c); to tie two ropes together, the square knot (d) or the fisherman's knot (f) (when they are the same size) or the sheet bend (e) (when one rope is thicker than the other). And there are many other knots adapted to these special tasks (see Figure P.2). Sailors use knots not only to moor boats, rig sails, and hoist loads, but also to make objects as varied as the regrettably famous "cat o' nine tails" and straw mats woven in Turk's head knots (Figure P.2b).

In the Age of Enlightenment (in England even earlier), oral transmission of maritime knot making was supplanted by specialized books about knots. One of the first authors in this genre was the Englishman John Smith, much better known for his romantic adventures with the beautiful Indian princess Pocahontas. At the same time, the terminology associated with knots became codified; it was even the subject of a detailed article in Diderot's and d'Alembert's *Enclopédie.*

Sailors were not the only inventors of knots. The fisherman's hook knot (Figure P.2f), the alpinist's chair knot (d), the engineer's constrictor knot (c), and the knitter's rice stitch (e) are only a few examples among many. The classic reference for knot making is Ashley's famous *Book of Knots* (1944). A few knots in particular derive from one of the greatest technological inventions of the Middle Ages: the pulley (Figure P.3a), together with the compound pulley (b and c). This work-saving device, a sort of Archimedes' lever with ropes, unites two major

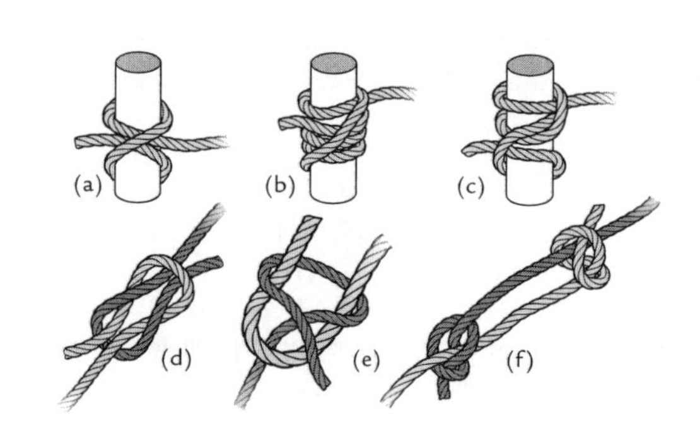

Figure P.1. Some sailors' knots.

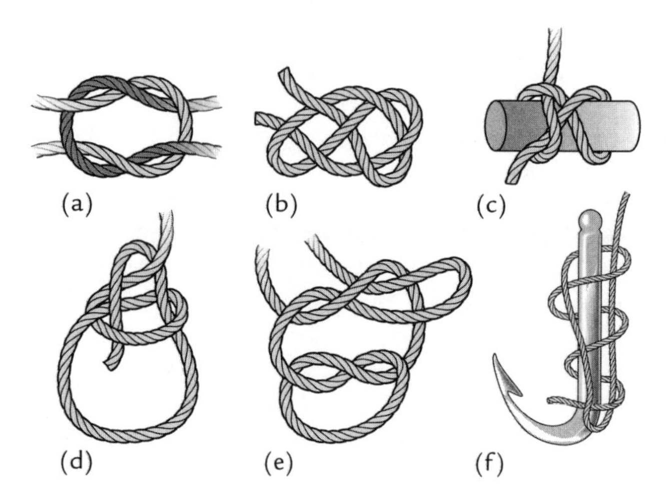

Figure P.2. Other knots.

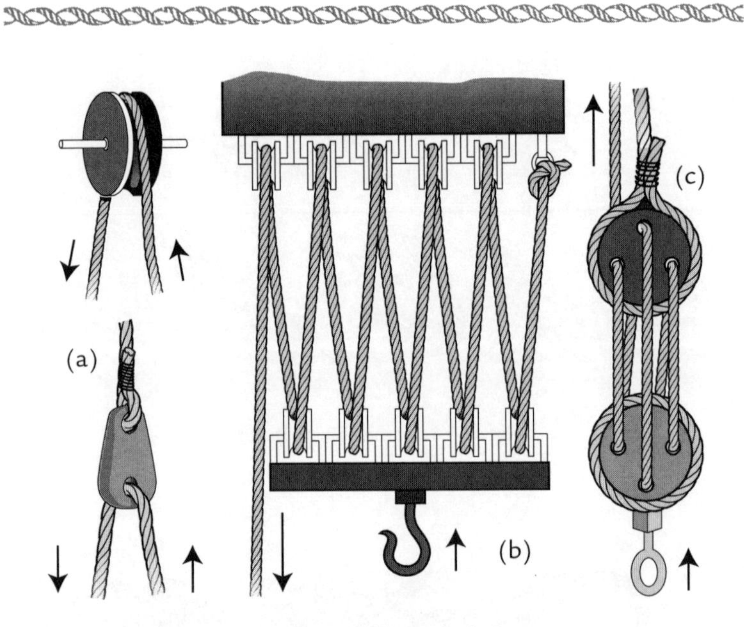

Figure P.3. Pullies and a hoist.

inventions of Antiquity: the wheel and the rope. It is used to pull or to lift all kinds of loads, usually also attached with the help of suitable knots. Thanks to knots, the rope became the universal technological tool of the age.

The technology for producing ropes (and cables) themselves—braiding—became very important. Fibers (once made of plants such as hemp, but synthetic in our times) had to be twisted into threads that were then braided into thicker strands, called lines, which in turn were braided in a specific way (generally involving three lines) to make a rope (see Figure P.4). The procedure for making cables is more com-

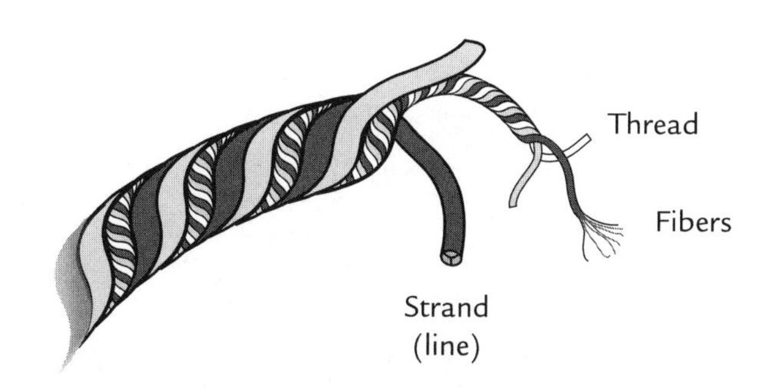

Figure P.4. Anatomy of a rope.

plex and involves four (or more) levels of cords, lines, and braided ropes. For the mathematician, the technology of braiding is the model for a basic idea in topology (as well as in mechanics)—the *braid*—which we will discuss in detail in Chapter 2.

Utilitarian and technological considerations aside, knots also have an aesthetic, mysterious, and magical aspect. As far as I know, it is precisely this feature of knots that is responsible for their first traces in our civilization. I have in mind the remarkable representations of knots on the megaliths and burial stones engraved by Neolithic peoples, in particular the Celts, during the fourth century B.C. Actually, these are chains of knots connected to one another (mathematicians call them *links*), as shown schematically in Figure P.5. We do not know the mystical and religious meaning of the links represented on menhirs (upright monuments also known as standing stones), but the geometric technology (based on regular figures) used to create these bewitching designs has been decoded by mathematicians (see Mercat, 1996).

Figure P.5. Links on a megalith.

Neolithic peoples were not alone in using links to decorate their objects of worship. Links are also found in the Middle Ages, in illuminated manuscripts, in the architecture of certain Eastern civilizations (friezes and other ornaments of the famous Alhambra palace in Spain are examples), and in the decorative elements framing icons in orthodox churches in northern Russia.

To end this overview of knots on a lighter note, think of the essential role they play in the magician's arsenal: knots that aren't, ropes that come undone instead of strangling the sexy magician's assistant, and so on. Some of these tricks (which amateur magicians can do) are described, from a mathematical vantage point, elsewhere (Prasolov and Sossinsky, 1997; Walker, 1985).

Let us move on to a summary of this book, to give a brief idea of what is to come and to allow those who don't intend to read the book from beginning to end to choose which chapters they wish to take in.[2] (Remember that the chapters are relatively independent.)

The first chapter has to do with the beginnings of the mathematical theory of knots, which was not the work of mathematicians—what a shame for them!—but that of physicists, more precisely, William Thomson (alias Lord Kelvin). The starting point (dating from around 1860) was Thomson's idea of using knots as models for the atom, models he dubbed "vortex atoms." To study the theory of matter from this point of view, he had to begin with knots. Fortunately for the self-esteem of mathematicians, Kelvin's theory ran aground and was soon forgotten, but not without leaving to posterity a series of problems (the *Tait conjectures*), which physicists were unable to solve at the time but mathematicians took care of a century later. The chapter not only deals with this spectacular failure of a beautiful physical theory, it also reviews various aspects of knot theory: Tait's tables of alternating knots, the superb *wild knots*, and *Antoine's necklace*. This last object provides us with an opportunity to talk about . . . blind geometers. The chapter ends with a brief discussion of the reasons for the failure of Thomson's theory.

The second chapter deals with the fundamental connection between knots and braids discovered by the American J. W. H. Alexander a half-century after Kelvin's abortive start. The mathematical theory of braids, which was formulated about the same time by the young German researcher Emil Artin, is more algebraic (and consequently simpler and more efficient) than knot theory. The connection in question (a geometric construction of childlike simplicity: the so-called closure of braids) enables one to obtain all knots from braids—Alexander's result. And because Artin rapidly established the classification of braids, it was natural to try to deduce the classification of knots from it. Efforts in this direction were unsuccessful, but they gave nice results, among which are the algorithms and software recently devised by French researchers.

In Chapter 3, I present a clever but simple geometric construction by the German mathematician Kurt Reidemeister, which reduces the study of knots in space to their planar projections (called *knot diagrams*). This gives us a chance to talk a little about catastrophe theory, encoding of knots, and working with knots on the computer. We will see that an algorithm invented by Reidemeister's compatriot Wolfgang Haken to determine whether a given knot can be untied does indeed exist, though it is very complex. That is because untying a knot often means first making it more complicated (alas, also true in real life). Finally, the functioning of an unknotting algorithm (which is fairly simple but has the disadvantage of futility when it comes to trying to unknot non-unknottable knots) will be explained: there, too, the modern computer does a better job of unknotting than we poor *Homo sapiens.*

Chapter 4 reviews the arithmetic of knots, whose principal theorem (the existence and uniqueness of prime knots) was demonstrated in

1949 by the German Horst Schubert. The curious resemblance between knots equipped with the composition operation (placing knots end to end) and positive integers (with the ordinary product operation) excited all sorts of hopes: Could knots turn out to be no more than a geometric coding of numbers? Could the classification of knots be just a plain enumeration? In Chapter 4 I explain why such hopes were unfulfilled.

Chapter 5 brings us to an invention that seems trite at first. It is due to the Anglo-American John Conway, one of the most original mathematicians of the twentieth century. As in Chapter 3, we will be dealing with small geometric operations carried out on knot diagrams. Contrary to Reidemeister moves, Conway operations can change not only the appearance but also the type of the knot; they can even transform knots into links. They make it possible to define and to calculate, in an elementary way, the so-called *Alexander-Conway polynomial*[3] of a knot (or link). These calculations provide a very easy and fairly efficient way to show that two knots are not of the same type, and in particular that some knots cannot be unknotted. But this method is probably not what the reader of this chapter will find most interesting: a biological digression explains how *topoisomerases* (recently discovered specialized enzymes) actually carry out Conway operations at the molecular level.

Chapter 6 presents the most famous of the knot invariants, the Jones polynomial, which gave new life to the theory fifteen years ago. In particular, it allowed several researchers to establish the first serious connections between this theory and physics. Oddly, it is the physical interpretation[4] of the Jones polynomial that gives a very simple description of the Jones invariants, whose original definition was far from elementary. This description is based on a tool—the Kauffman

bracket—that is very simple but that plays no less fundamental a role in modern theoretical physics. This chapter contains several digressions. In one of them, readers will learn that the main ingredient in the Kauffman bracket was already known in the Neolithic age by the Celtic artists mentioned earlier.

Chapter 7 is devoted to the last great invention of knot theory, Vassiliev invariants. Here, too, the original definition, which drew on catastrophe theory and spectral sequences,[5] was very sophisticated, but an elementary description is proposed. Instead of complicated mathematical formulas, readers will find abbreviated calculations involving little diagrams, along with a digression on the sociological approach to mathematics.

The eighth and final chapter discusses connections between knot theory and physics. Contrary to what I tried to do in the other chapters, here I could only sketch out the most rudimentary explanations of what is going on in this area. I had to use some new technical terms from mathematical physics without being able to explain them properly. But I am convinced that even readers closer to the humanities than to the sciences will succeed in getting through this chapter. Even if they cannot grasp the precise meaning of the terms and equations, they can focus on the gist of the discussion, on the role of coincidences, and on the dramatic and emotional side of contemporary research.

The brilliant beginnings of knot theory, over 130 years ago, were marked by a ringing failure—as a physical theory of matter—but the concepts were revived thanks to the repeated efforts of mathemati-

cians, whose only motivation was intellectual curiosity. Progress required new, concrete ideas. And the ideas came, springing from the imagination of the best researchers, often sparking exaggerated hopes. But every failure made it easier to grasp the remaining problems, making the final goal ever more attractive. Today we are in a situation similar to that of 1860: some researchers think, as William Thomson did, that knots play a key role in the basic theory of the structure of matter. But that is not to say that we are back at the beginning: the spiral of knowledge has made a full loop, and we find ourselves at a higher level.

The theory of knots remains just as mysterious and vibrant as ever. Its major problems are still unsolved: knots continue to elude efforts to classify them effectively, and still no one knows whether they possess a complete system of invariants that would be easy to calculate. Finally, the basic role knots are supposed to play in physics has not yet been specified in a convincing way.

KNOTS

1

ATOMS AND KNOTS

(Lord Kelvin · 1860)

In 1860, the English physicist William Thomson (better known today by the name of Lord Kelvin, but at the time not yet graced with a noble title) was pondering the fundamental problems linked to the structure of matter. His peers were divided into two enemy camps: those who supported the so-called corpuscular theory, according to which matter is constituted of *atoms,* rigid little bodies that occupy a precise position in space, and those who felt matter to be a superposition of *waves* dispersed in space-time. Each of these theories provides convincing explanations for certain phenomena but is inadequate for others. Thomson was looking for a way to combine them.

And he found one. According to Thomson, matter is indeed constituted of atoms. These "vortex atoms," however, are not pointlike objects but little knots (see Thomson, 1867). Thus an atom is like a wave that, instead of dispersing in all directions, propagates as a narrow beam bending sharply back on itself—like a snake biting its tail. But this snake could wriggle in a fairly complicated way before biting itself, thus forming a knot (Figure 1.1). The type of knot would then determine the physicochemical properties of the atom. According to this

view, molecules are constituted of several intertwined vortex atoms, that is, they are modeled on what mathematicians call *links:* a set of curves in space that can knot up individually as well as with each other.

This theory will no doubt seem rather fanciful to the reader accustomed to Niels Bohr's planetary model of the atom taught in school. But we are in 1860, the future Nobel laureate will not be born until 25 years later, and the scientific community is taking Thomson's revolutionary idea seriously. The greatest physicist of the period, James Clerk Maxwell, whose famous equations formed the basis of wave theory, hesitated at first, then warmed to the idea. He insisted that Thomson's theory explained the experimental data accumulated by researchers better than any other.

To develop his theory, Thomson needed first of all to see which different types of knots are possible; in other words, he had to classify knots. It would then have been possible to classify atoms by associating each type of knot with a specific atom. For example, the three

Figure 1.1. Model of an atom?

  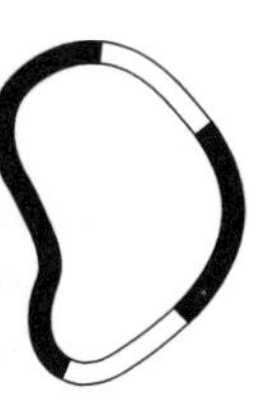

Figure 1.2. Three knots: the trefoil, the figure eight, and the unknot.

knots represented in Figure 1.2, the *trefoil,* the *figure eight,* and the *trivial knot* or *unknot,* could be models of carbon, oxygen, and hydrogen, respectively.

So in the beginning the problem was mathematical (rather than physicochemical): the problem of classifying knots. And it was a Scottish physicist and mathematician, a friend of Thomson, Peter Guthrie Tait, who set out to solve it.

Tait, Kirkman, and the First Tables of Knots

According to Tait, a knot, being a closed curve in space, could be represented by a planar curve obtained by projecting it perpendicularly on the horizontal plane. This projection could have *crossings* (Figure 1.3), where the projection of one part of the curve crossed another; the planar representation shows the position in space of the two strands that cross each other by interrupting the line that represents the lower strand at the crossing. I have already used this natural way of drawing knots (Figure 1.2) and will continue to use it.

Posing the question of how knots should be classified requires

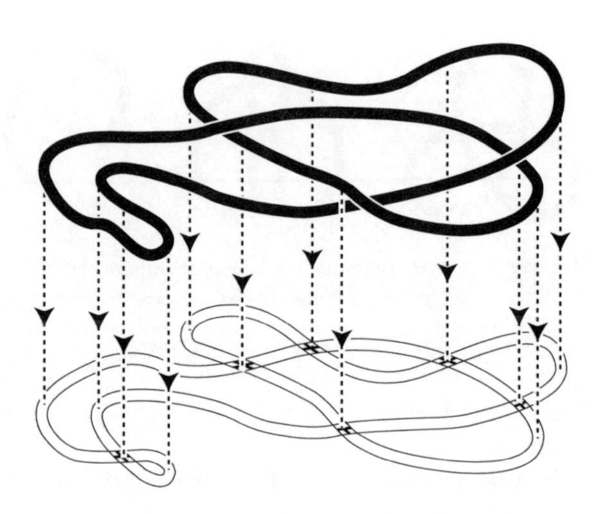

Figure 1.3. Planar projection of a knot.

specifying which knots belong to the same class. It requires, in other words, *precisely defining the equivalence of knots.* But we will leave this definition (the *ambient isotopy* of knots) for later, limiting ourselves here to an intuitive description. Imagine that the curve defining the knot is a fine thread, flexible and elastic, that can be twisted and moved in a continuous way in space (cutting and gluing back is not allowed). All possible positions will thus be those of the same knot.

Changing the position of the curve that defines a knot in space by moving it in a continuous way (without ever cutting or retying it) always results in the same knot by definition, but its planar representation may become unrecognizable. In particular, the number of crossings may change. Nevertheless, the natural approach to classifying

Figure 1.4. Two representations of the same knot.

knots in space consists first of making a list of all the planar curves with 1, 2, 3, 4, 5, . . . crossings, then eliminating the duplications from the list, that is, the curves that represent the same knot in space (Figure 1.4).

Of course, for the task to be doable within a human lifetime, the maximum number of crossings of the knots considered has to be limited. Peter Tait stopped at 10. Tait had an initial stroke of luck: he learned that an amateur mathematician, the Reverend Thomas Kirkman, had already classified planar curves with minimal crossings, and all that remained was to eliminate the duplications systematically. But that is not so easy to do. Indeed, for each crossing of a planar curve, there are two ways to decide which strand in the crossing should be uppermost. For a curve with 10 crossings, for example, a priori there are 2^{10}, or 1,024, possibilities for making a knot. Tait decided to list only *alternating knots,* that is, those in which overpasses and underpasses alternate along the curve (Figure 1.5). In this way, exactly two

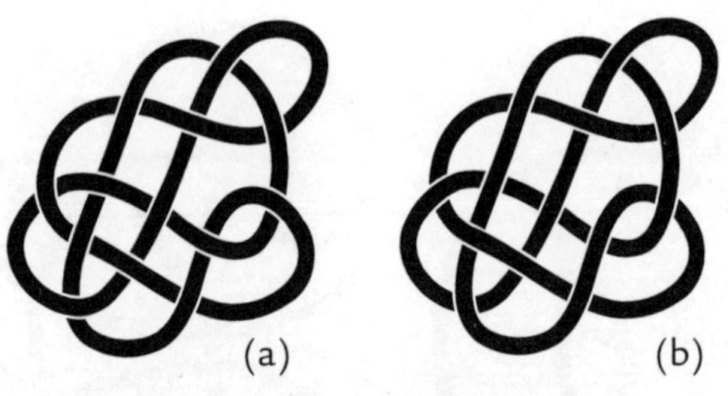

Figure 1.5. An alternating knot (a) and a nonalternating knot (b).

alternating knots corresponded to each planar curve, substantially facilitating Tait's task. Which is not to say that it became simple: he devoted the rest of his life to it.

Nonalternating knots (one with 10 or fewer crossings) were classified in 1899 by C. N. Little, after six years of work. Little managed to avoid the systematic run-through of the 2^{10} uncrossing possibilities (for each knot) mentioned above. Unfortunately for Thomson, Kirkman, Little, and Tait, by the time Little and Tait finished their work, almost no one was interested in knot tables for reasons that will be explained at the end of this chapter.

Still, at the century's close, most of the work on classifying knots (with 10 or fewer crossings) had been done, and tables of knots appeared. Figure 1.6 shows an example, a table of (prime) knots with 7 or fewer crossings. The exact meaning of the expression "prime knot," which is analogous to "prime number" in the sense that it cannot be factored, is explained in Chapter 4, which deals with the arithmetic of knots. But before continuing this account of Kelvin's and Tait's work, it is worth making a few points about classifying knots.

Figure 1.6. Table of knots with seven or fewer crossings.

A Mathematical View of Knot Classification

Let us pose the problem in precise terms, rigorous enough to satisfy a mathematician (readers disinclined to scientific rigor can skip this part after glancing at the figures). First of all, the very concept of a knot needs to be defined. We define a knot, or more precisely *a representation of a knot,* as a closed polygonal curve in space (Figure 1.7a). A *knot* is just a class of equivalent representations of knots, equivalence being the relation of ambient isotopy, defined as follows: An *elementary isotopy* is achieved either by adding a triangle (as we add *ABC* in Figure 1.7b) to a segment (*AB*) of the polygonal curve, then replacing this segment by the two other sides (*AC* and *CB*), or by doing the opposite. Of course, the triangle must have no points in common with the polygonal curve other than its sides. An *ambient isotopy* is just a finite sequence of elementary isotopies (Figure 1.7c).

Clearly this definition corresponds to our sense of a knot as the abstraction of a string whose ends are stuck together, and ambient isotopy allows us to twist and move the knot in space as we do with a real string (without tearing it). From the aesthetic point of view, it is perhaps not satisfying to think of strings that have angles everywhere, but it is the price to pay for defining a knot in a manner both elementary and rigorous.[1]

Representing a knot as a polygonal curve has a motive other than the ability to attach triangles to it (which presupposes that the "curve" is made of segments); in fact, it is also a necessary condition for avoiding "local pathologies." It is required to avoid the so-called *wild knots,* which are not topologically equivalent to polygonal curves (or to a smooth curve). Wild knots are the result of a process of infinite knot-

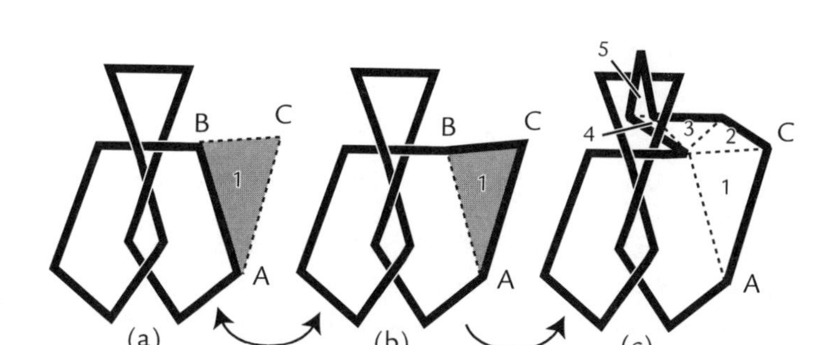

Figure 1.7. A knot drawn as a closed polygonal curve in space (a) and two equivalent representations (b and c) of the same knot (isotopy).

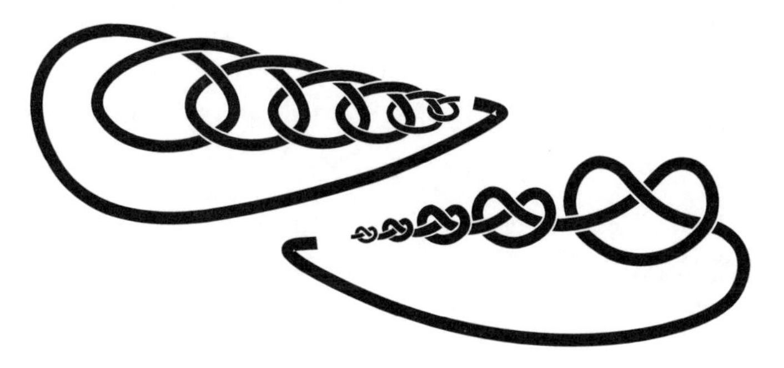

Figure 1.8. Wild knots.

ting: the twists of the curve get smaller and smaller and converge toward a limit point known as a *wild point* of the curve (Figure 1.8).

Rigorously defining a knot (as a polygonal curve or a smooth curve) makes it possible to avoid these little horrors and simplifies the theory. Before we continue our preliminary investigation of "tame" knots, here are a few remarks on their "wild" kin (with a few drawings included).

Digression: Wild Knots, Spatial Intuition, and Blindness

The examples of wild knots shown up to now possess a single isolated pathological point, toward which a succession of smaller and smaller knots converge. Wild knots with several points of the same type can easily be constructed. But one can go further: Figure 1.9 shows a wild knot that has an infinite (even uncountable, for those who know the expression) set of pathological points.

This set of wild points is in fact the famous *Cantor continuum,* the set of points in the segment [0, 1] that remain after one successively eliminates the central subinterval (1/3, 2/3), then the (smaller) central subintervals (1/9, 2/9) and (7/9, 8/9) of the two remaining segments,

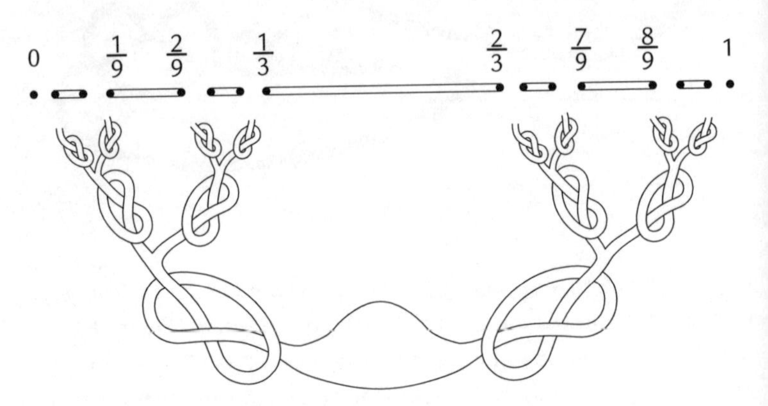

Figure 1.9. A wild knot converging to Cantor's continuum.

then the four (tiny) central subintervals (1/27, 2/27), (7/27, 8/27), (19/27, 20/27), and (25/27, 26/27) of the four remaining segments, and so on to infinity.

A much more beguiling wild knot can be obtained by making a curve pass through a set even more complicated than Cantor's continuum, for example, Antoine's necklace. No, I do not mean a gift of the Roman general[2] to Cleopatra, but a geometric construction devised by the French mathematician Louis Antoine. Let us try to describe this jewel of the mathematical imagination, shown in Figure 1.10.

Begin with a solid torus, T_1, in the shape of a doughnut and place inside it four thinner solid tori linked two by two to make a chain, T_2, of four rings. Inside each of these four rings of the chain T_2, construct a chain similar to the preceding one; the set formed by these four little chains (constituting the 16 tiny tori) is denoted T_3. Inside each tiny torus knot, take . . . The process continues indefinitely, and the set obtained as the infinite intersection of the sets T_i will be Antoine's necklace:

$$A = T_1 \cap T_2 \cap \ldots \cap T_n \cap \ldots$$

Antoine's necklace has some remarkable properties that I shall not dwell on: it will simply aid us to construct a wild knot invented by the Russian mathematician G. Ya. Zuev and represented in the same figure. The knot in question is shown (partially) in the form of the curve that insinuates itself inside the large solid torus, then inside the smaller tori, then into the tiny tori, and so on, dividing each time it penetrates a torus knot to tend toward Antoine's necklace. One can show (but the rigorous demonstration is rather tricky) that the curve

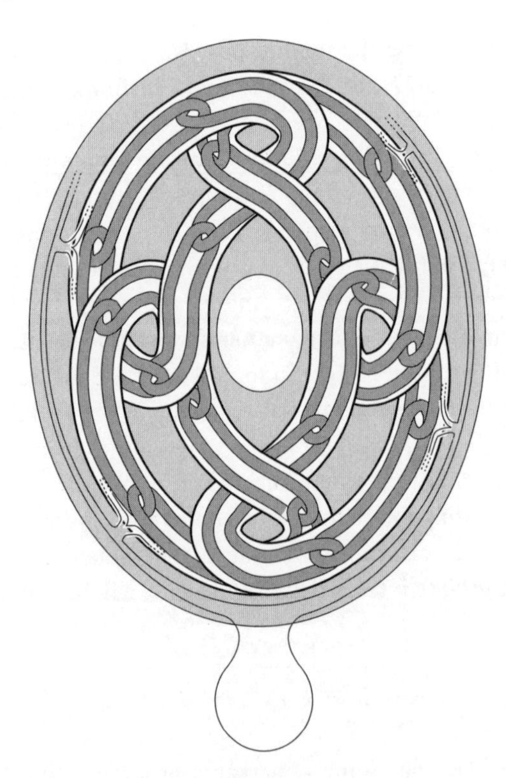

Figure 1.10. A wild knot converging to Antoine's necklace.

ultimately obtained is indeed a simple closed curve, and the set of its wild points is precisely Antoine's necklace.

Inventing monsters such as Antoine's necklace or Zuev's wild knot takes a considerable ability to visualize three-dimensional space. So it might surprise the reader to learn that these two mathematicians are

blind. But actually, it isn't all that surprising, since almost all blind mathematicians are (or were) geometers. The spatial intuition that sighted people have is based on the image of the world that is projected onto their retinas; thus it is a two- (and not three-) dimensional image that is analyzed in the brain of a sighted person. A blind person's spatial intuition, on the other hand, is primarily the result of tactile and operational experience. It is also deeper—in the literal as well as the metaphorical sense.

As we leave this digression, note that recent biomathematical studies (based on work with children and adults who were born blind but gained their sight afterward) have shown that the deepest mathematical structures, such as topological structures, are innate, whereas finer structures, such as linear structures, are acquired. Thus, at first, the blind person who regains his sight does not distinguish a square from a circle: he sees only their topological equivalence. In contrast, he immediately sees that the torus is not a sphere. As for us, our tendency to consider what we see the "absolute truth" often makes us conceive the world in a very flat and superficial way.

The Failure of Thomson's Theory

While European physicists were debating the merits of Thomson's theory and Tait was filling in his tables of knots, another researcher—unknown and working in an immense, underdeveloped country—was, like Thomson and Tait, reflecting on the structure of matter. He, too, was trying to establish tables of atoms but, little inclined to geometric considerations, he based his tables on arithmetical relationships among the various properties of chemical elements.

This scientist made an unexpected discovery: there are very simple,

until then unnoticed, relationships among the chemical properties. And he published what today is called the *periodic table of elements.* It would take some time for this remarkable discovery to be recognized in western Europe. Once that happened, my compatriot Mendeleev buried the idea of atoms as knots. Thomson's theory had not done much for chemistry, and it was quickly replaced by the arithmetical theory of Mendeleev. And physicists, ashamed and embarrassed, forgot about knots for nearly a century. It was the mathematicians who would take up the subject once again.

2

BRAIDED KNOTS

(Alexander · 1923)

This chapter is devoted to a remarkable connection, discovered by mathematicians, between two beautiful topological objects: braids and knots. Mathematically speaking, what is a braid? Roughly, it is the formal abstraction of what is meant by a braid in everyday language (a braid in a young girl's hair, a plaited key chain, a braided dog's leash, a classic twisted rope, and so on)—in other words, some strings tangled in a certain way. More precisely, you can imagine a *braid* of n *strands* as n threads attached "above" (to horizontally aligned nails) and hanging "down," crossing each other without ever going back up; at the bottom, the same threads are also attached to nails, but not necessarily in the same order (Figure 2.1).

The strands of a braid can be rearranged (without detaching the top and bottom, and of course without tearing or reattaching them) to get a braid that looks different but is *equivalent* to (or an isotope of) the first braid (Figure 2.2). As with knots, we do not distinguish two isotopic braids: we think of them as two representations of the same object (from the formal mathematical point of view, this means that the ob-

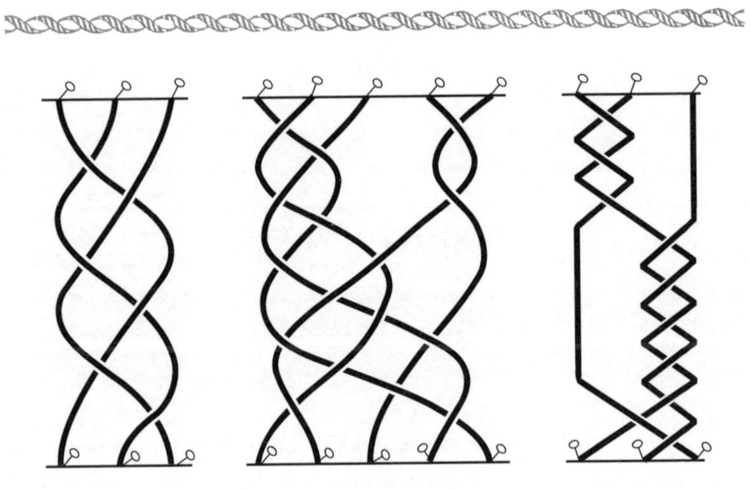

Figure 2.1. Examples of braids.

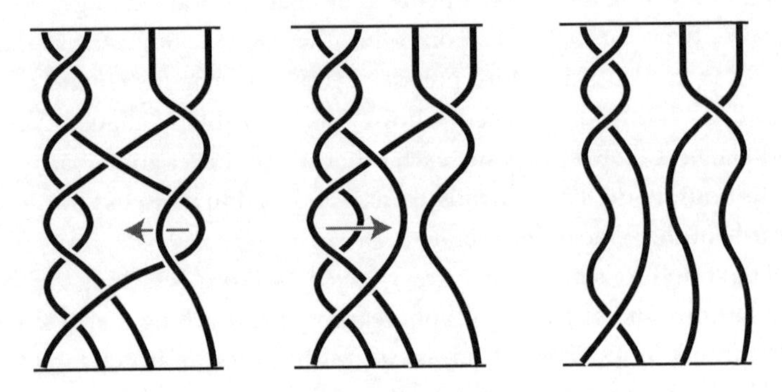

Figure 2.2. Isotopes of a four-stranded braid.

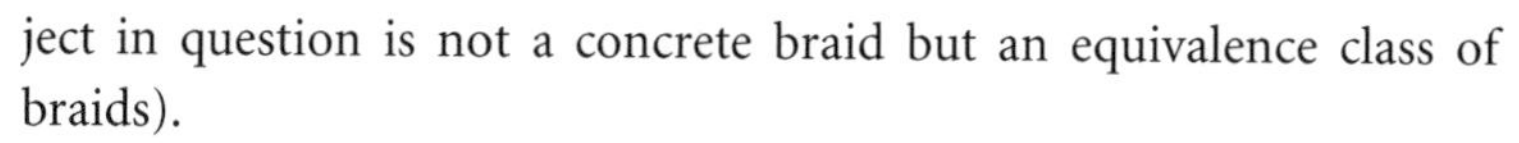

ject in question is not a concrete braid but an equivalence class of braids).

The basics of braid theory were developed by Emil Artin in the 1920s. The theory is a marvelous blend of geometry, algebra, and algorithmic methods. It has a bunch of applications ranging from the textile industry to quantum mechanics by way of topics as varied as complex analysis, the representation of functions of several variables as the composition of functions of a lesser number of variables, and combinatorics. But we will deal with this theory later, since our immediate goal is to understand the connection between braids and knots.

Closure of Braids

A knot can be made from a braid by the operation of *closure*, which means joining the upper ends of the strands to the lower ends (see Figure 2.3a).

Will a knot always be obtained in this way? According to Figure 2.3b, not always: the closure of a braid may very well result in a link of several components (that is, several curves, in contrast to a knot, which by definition consists of only one curve). An attentive reader of the previous chapter will recognize the trefoil knot in Figure 2.3a—though perhaps not at once.

The following question arises immediately: Which knots can be obtained in this way? The answer, found by J. W. H. Alexander in 1923, explains the importance of braids in knot theory: they all can. Alexander's theorem can be expressed as follows: *Every knot can be represented as a closed braid.* (Actually, Alexander showed that this assertion is true for the more general case of links, of which knots are just an example.)

Alexander probably hoped that his theorem would be a decisive step forward in classifying knots. Indeed, as we will see later, braids are much simpler objects than knots; the set of braids possesses a very clear algebraic structure that enables one to classify them. Is it reasonable, then, to try to use braids to classify knots? How did this idea develop? We will see at the end of the chapter.

For now, let us return to Alexander's theorem: How can it be proved? Given a knot, how does one find the braid whose closure would be that knot? First of all, note that the desired braid can easily be seen when the braid is rolled up in a coil, that is, when it always turns in the same direction around a certain point (as the knot in Fig-

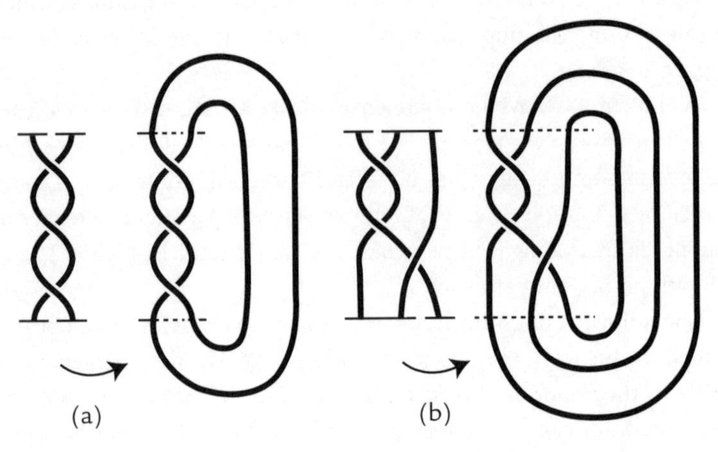

Figure 2.3. Closure of two braids.

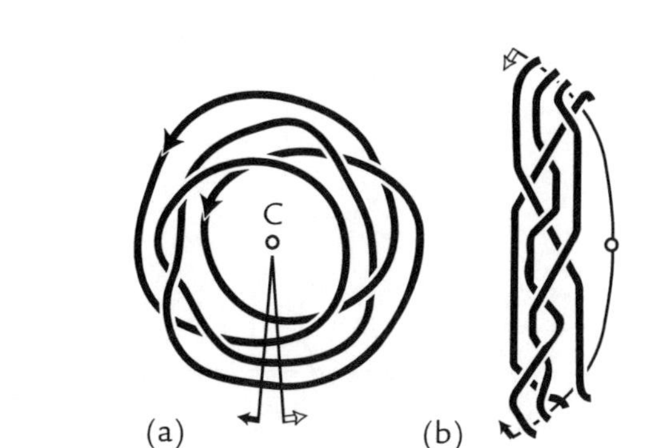

Figure 2.4. Unrolling a coiled knot into a braid.

ure 2.4 turns around the center *C*). In this situation, all that is required to find the braid is to cut the knot along a line extending outward from the center and then to unroll it (Figure 2.4b).

But what if the knot is not coiled, for example, the knot shown in Figure 2.5a? (As readers of the first chapter will know, this knot is called a figure eight knot.) In that case, just move the "fat" part (the thicker line) of the knot (the one "going the wrong way") over the point *C* on the other side of the curve. The resulting coiled knot (Figure 2.5b) can then be unrolled into a braid as in the preceding example (Figure 2.5c).

Actually, this elegant method (transforming any knot into a coiled knot) is universal, and it allowed Alexander to prove his theorem. Its weakness—and there is one—is its ineffectiveness from a practical point of view; specifically, it is difficult to teach to a computer. Another method of braiding knots, more doable and easier to program,

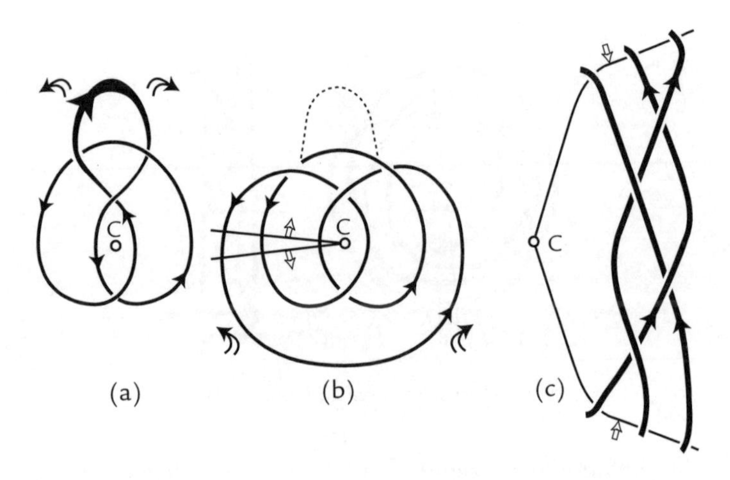

Figure 2.5. Coiling a figure eight knot and unrolling it into a braid.

was invented by the French mathematician Pierre Vogel. The reader not inclined to algorithmic reasoning can blithely skip this description and go on to the study (much simpler and more important) of the group of braids.

Vogel's Braiding Algorithm

The braiding algorithm transforms any knot into a coiled knot. To describe it I must introduce some definitions related to the planar representations of knots. Assume that a given knot is oriented; that is to say, the direction of the curve (indicated by arrows) has been selected. The planar representation of the knot defines a kind of geographic map in

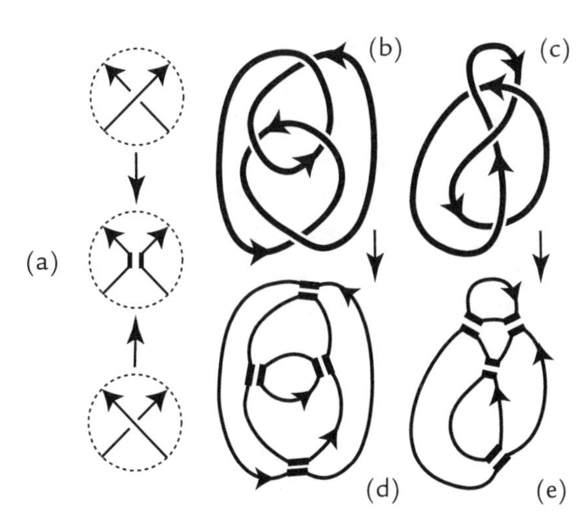

Figure 2.6. Desingularized knots and Seifert circles.

the plane, the regions or *countries* being the areas bounded by parts of the knot's curve. On this map, the border of each country consists of several edges (oriented in accordance to the arrows) that join one crossing of the knot to a neighboring crossing. Included among the countries is the infinite region—the one located outside the curve.

Because the curve of the knot is oriented, the crossings are marked with arrows; these make it possible to "desingularize" the knot N unambiguously, in other words, to replace all the crossings of N by their *smoothings* as shown in Figure 2.6a. Desingularization transforms each knot into one or more oriented, closed curves (without crossings), called *Seifert circles,* that represent the knot (Figures

2.6d, 2.6e). Two Seifert circles are *nested* if one of them is inside the other and if the orientations of the two circles coincide. Note that desingularizing a coiled knot always yields a nested system of Seifert circles, and vice versa (Figure 2.6b).

On the other hand, when Seifert circles are not nested (as in Figure 2.7b), the change-of-infinity[1] operation nests the two circles (Figure 2.7c). In fact, this figure shows that while circles 1 and 2 are nested, circle 3 does not encompass them; but inverting this last circle results in the circle 3′, which neatly encompasses circles 1 and 2. (In this case, the change-of-infinity move resembles the operation carried out in Figure 2.5.)

Let us now consider the planar map determined by the knot *N*. A country in this map is said to be *in turmoil* if it has two edges that belong to two different Seifert circles, labeled with arrows going in the same direction around the region. For example, in the smoothing of knot *N* in Figure 2.8, the country *H* is in turmoil, whereas regions P_1 and P_2 are not: since the thick edges head in the same direction around *H* and belong to two distinct Seifert circles, *H* is in turmoil; P_1 is not, because its edges belong to a single Seifert circle; finally, P_2 is not in turmoil either, because its edges go around P_2 in opposite directions.

An operation called *perestroika* can be applied to any country in turmoil (Figure 2.9). *Perestroika* consists in replacing the two faulty edges by two "tongues," one of which passes over the other, forming two new crossings. The result is to create a central country (not in turmoil) and several new countries, some of which (in this case, two) may be swallowed up by bordering countries. I am sure that now the reader understands the choice of the geopolitical term *perestroika*.

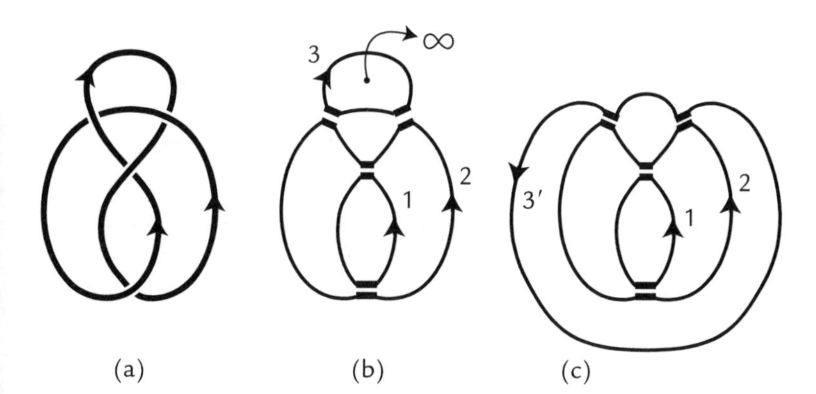

Figure 2.7. Change of infinity.

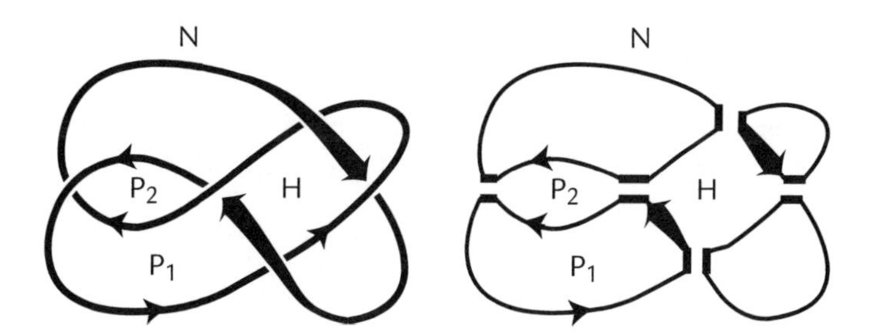

Figure 2.8. Countries in and out of turmoil.

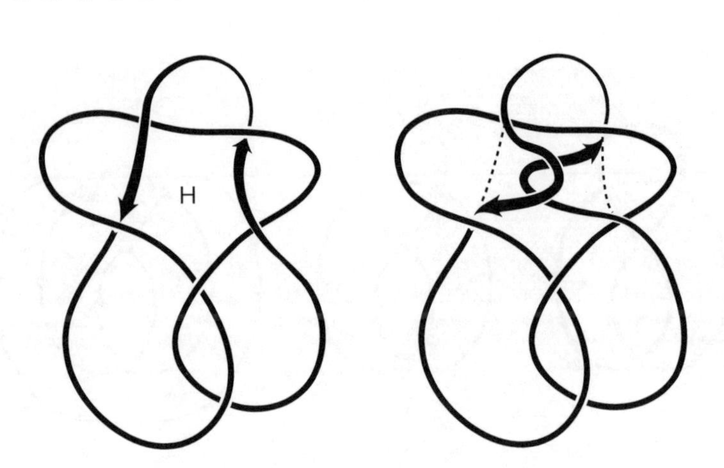

Figure 2.9. *Perestroika* of a country in turmoil.

Vogel's algorithm can now be presented in the form of a "program" written in a sort of "pseudo-Pascal":

```
Do smoothing

While: There is a disjoint region
    Do perestroika
    Do smoothing
End while

While: The Seifert circles are not nested
    Do change infinity
End while
Stop
```

Most of this language was explained above, but the command Do change infinity requires clarification: it means taking one of the small-

est Seifert circles not nested with the others and sending a point inside this circle to infinity.

First let us apply the Vogel algorithm to a very simple knot (the unknot, in fact) to see how changes of infinity occur (Figure 2.10). Following the first smoothing, there are no countries in turmoil, and no nested Seifert circles. So we proceed to the command `Do change infinity`, which must be carried out twice—(b) becomes (c) and (c) becomes (d)—to obtain a coiled knot (d), which can then be unrolled into a braid as before (e).

Figure 2.11 shows how Vogel's algorithm coils a knot with five crossings.[2] Following the initial smoothing, the loop (in the computational sense of the word) contains two perestroikas (Figures 2.11b and 2.11c); it is followed by a change-of-infinity move. The result (Figure 2.11d) is indeed a coiled knot, even if it does not look like one. To make sure, I have redrawn it twice (Figures 2.12b, 2.12c). The reader will have no difficulty recognizing the rolled-up knot from Figure 2.4a (and so can admire the desired braid, which appeared earlier as Figure 2.4b).

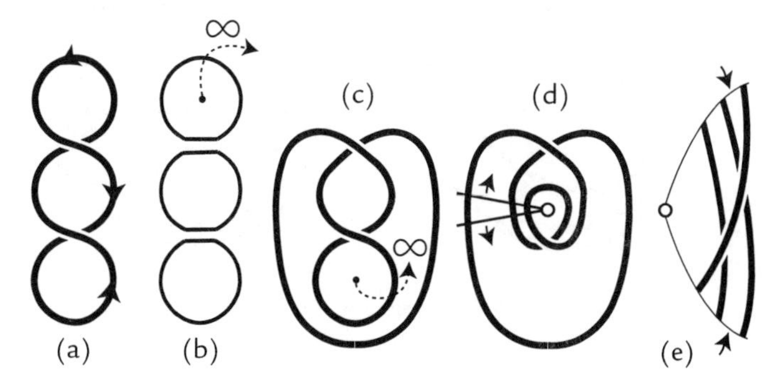

Figure 2.10. Vogel algorithm applied to the unknot.

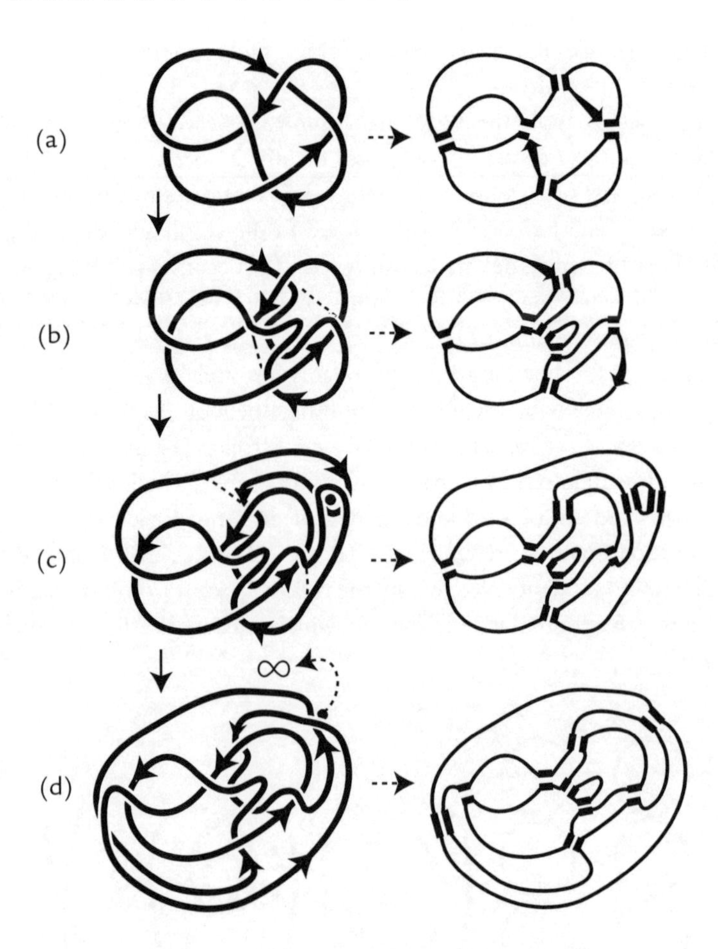

Figure 2.11. Vogel algorithm applied to knot 5^2.

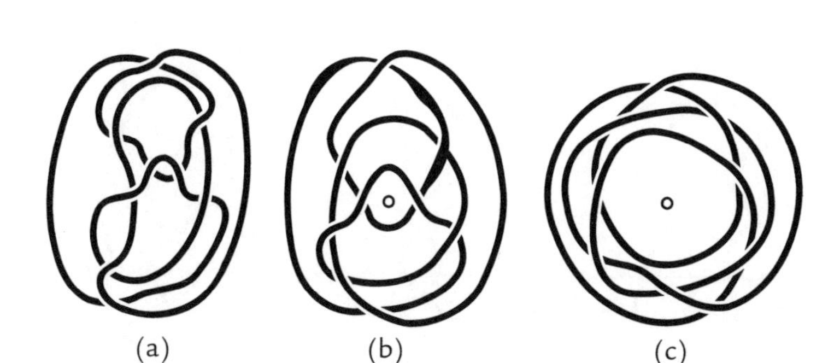

Figure 2.12. Using the Vogel algorithm to unroll a knot.

Note that it is not at all obvious that the algorithm, which includes two While loops that are a priori dangerous, will always terminate. Yet it does, and very rapidly. Proving that the second loop always terminates is elementary. On the other hand, to prove that the same is true for the first loop, Vogel had to use fairly sophisticated algebraic topology methods.

Of course, to transform our "program" into software for a real computer, we have to know how to code knot representations in such a way that the machine will be able to work with them. We will come back to the coding of knots in Chapter 4.

The Braid Group

Let us return to the study of braids. First of all, we are going to define an operation, the *composition* or *product*, on the set of braids with n

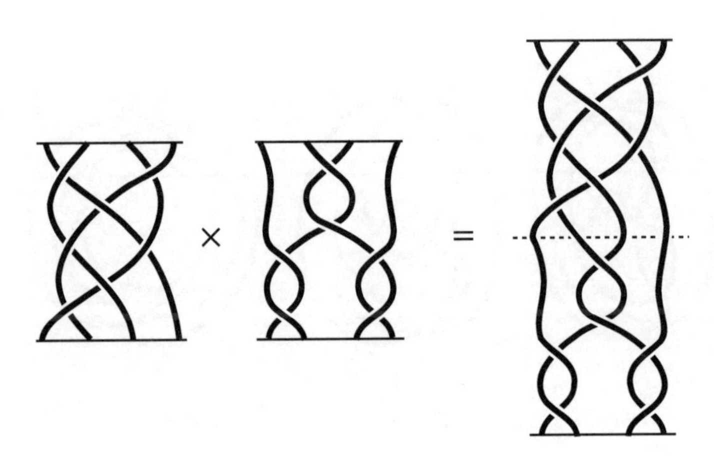

Figure 2.13. The product of two braids.

strands. This operation consists simply of placing the braids end to end (by joining the upper part of the second braid to the lower part of the first), as in Figure 2.13.

It turns out that the product of braids possesses several properties that resemble the ordinary product of numbers. First, there is a braid known as the *unit braid* (e), a braid that, like the number 1, does not change what it multiplies. This is the *trivial braid,* whose strands hang vertically without crossing. Sure enough, appending a trivial braid to a given braid amounts to extending its strands, which does not change the class of the braid in any way.

Second, for each braid b there exists an *inverse braid,* b^{-1}, whose product with b gives the trivial braid: $b \cdot b^{-1} = e$ (just as for each number n, its product with the inverse number $n^{-1} = 1/n$ is equal to one, $n \cdot n^{-1} = 1$). As can be seen in Figure 2.14, the inverse braid is the

braid obtained by taking the horizontal mirror image of the given braid; indeed, each crossing cancels with its mirror image, such that all the crossings gradually dissolve, two by two, beginning at the middle of the product braid.

The third property common to braids and to numbers is the *associativity* of the product operation: $(a \cdot b) \cdot c = a \cdot (b \cdot c)$ is always true. When a set is endowed with an operation that enjoys all three of the properties just described, mathematicians call this set a *group*. Thus, I have just shown that braids with n strands form a group. This group will be denoted by B_n.

In contrast to numbers, however, the braid group B_n (for $n > 2$) is not commutative: the product of two braids generally depends on the order of the factors.

The product of braids makes it possible to replace the picture repre-

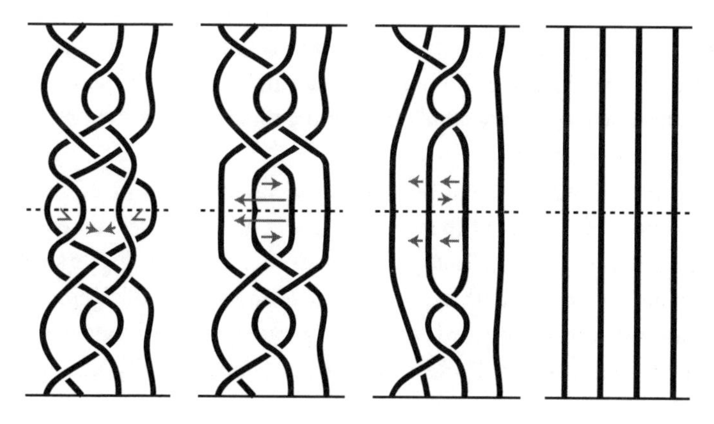

Figure 2.14. The product of a braid and its inverse.

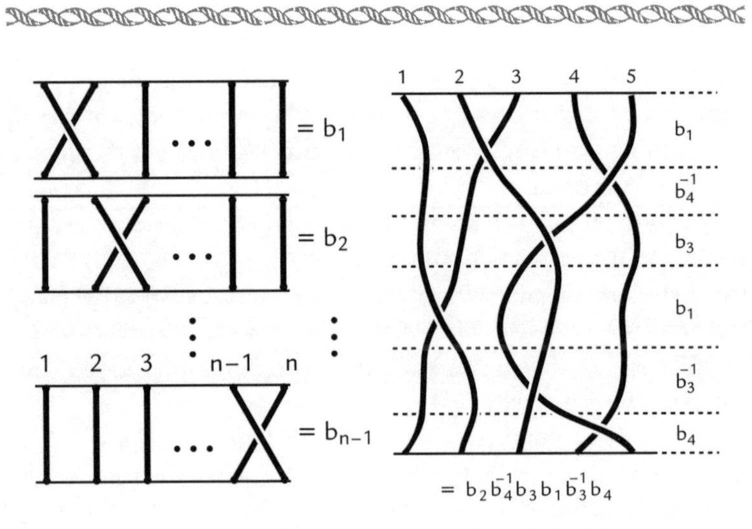

Figure 2.15. Algebraic representation of a braid.

senting a braid by a word—the algebraic encoding of that braid. Indeed, moving along a braid from top to bottom, we see that it is the successive product of braids each with a single crossing (Figure 2.15); we call these *elementary braids* and denote them by $b_1, b_2, \ldots, b_{n-1}$ (for braids with n strands).

So we have replaced braids—geometric objects—by words: their algebraic codes. But recall that the geometric braids possess an equivalence relation, namely, isotopy. What does that mean algebraically? Artin had an answer to this question. He found a series of algebraic relations between braid words that gave an adequate algebraic description of their isotopy. These relations are commutativity for distant braids

$$b_ib_j = b_jb_i \text{ for } \mid i - j \mid \geq 2, \qquad i, j = 1, 2, \ldots, n - 1$$

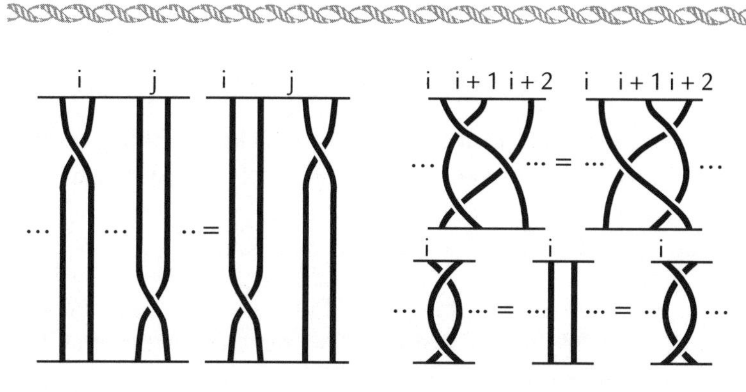

Figure 2.16. Relations among the group of braids.

and Artin's relation (or the *braid relation*)

$$b_i b_{i+1} b_i = b_{i+1} b_i b_{i+1}, \qquad i = 1, 2, \ldots, n-2$$

Their geometric interpretation is shown in Figure 2.16. To an observer with a little spatial imagination, it is immediately obvious that these relations are valid for braids, that is, they do indeed correspond to isotopies.

What is less obvious—and is one of Artin's key findings—is that these two relations (if one adds the trivial relations,[3] $b_i b_i^{-1} = e = b_i^{-1} b_i$, also shown in Figure 2.16) suffice to replace the geometric manipulations related to isotopy by admissible algebraic calculations on the braid words; each of the admissible calculations consists in replacing a part of a word identical to one of the members of the relations that appear in Figure 2.16 by the other member of that relation. Here is an example of an admissible calculation in the group of braids with four strands, B_4:

$$b_3^{-1}(b_2b_3b_2)b_3^{-1} = (b_3^{-1}b_3)\ b_2\ (b_3b_3^{-1}) = eb_2e = b_2$$

(To make this formula easier to read, I have enclosed within parentheses the parts of the word that are replaced successively during the calculation.)

More precisely, Artin's theorem affirms that

> Two braids are isotopic if and only if the word representing one of them can be transformed into the word representing the other by a sequence of admissible calculations.

Artin's theorem is important because it reduces the geometric study of braids to their algebraic study, which is generally more efficient. This algebraic approach allowed Artin to classify braids. Put in other words, it allowed him to find a comparison algorithm for them; for each pair of braids, the algorithm says "no" if they are not isotopic and "yes" if they are (as well as providing a set of admissible calculations that lead from one to the other in the latter case).

Classifying Braids

I will not be describing any of the braid-comparison algorithms here, neither Artin's (which he called by the lovely word *combing*), nor the one recently discovered by the French mathematician Patrick Dehornoy, which is much simpler and more efficient. But to convince you of the efficiency of algebraic-algorithmic methods in geometry, I chose (more or less at random) an example of a calculation carried out by my little computer (which has in the recesses of its electronic memory some software that drives Dehornoy's algorithm). This calcu-

lation, which takes place in the group B_4 and uses the (more readable) notation *a, A, b, B, c, C* for the basic braids b_1, b_1^{-1}, . . . , b_3^{-1}, respectively, shows that a braid with four strands that looks rather complicated is in fact the trivial braid.

ABBAAAAA[Abbbbbbbcba]AccBCaBBBBBBaaaaaaBB
= *ABBAAAAA[Aba]aaaaaaaBcbaBAccBCaBBBBBBaaaaaaBB*
= *ABBAAAA[Aba]BaaaaaaaBcbaBAccBCaBBBBBBaaaaaaBB*
= *ABBAAA[Aba]BBBaaaaaaaBcbaBAccBCaBBBBBBaaaaaaBB*
= *ABBAA[Aba]BBBaaaaaaaBcbaBAcBCaBBBBBBaaaaaaBB*
= *ABBA[Aba]BBBBaaaaaaaBcbaBAcBCaBBBBBBaaaaaaBB*
= *ABB[Aba]BBBBBaaaaaaaBcbaBAccBCaBBBBBBaaaaaaBB*
= *[ABa]BBBBBBaaaaaaaBcbaBAccBCaBBBBBBaaaaaaBB*
= *[ABa]BBBBBBaaaaaaaBcbaBAccBCaBBBBBBaaaaaaBB*
= *b[ABBBBBBBBa]aaaaaaaBcbaBAccBCaBBBBBBaaaaaaBB*
= *bbAAAAAA[Aba]aaaaaBcbaBAccBCaBBBBBBaaaaaaBB*
= *bbAAAAAAb[Aba]aaaaBcbaBAccBCaBBBBBBaaaaaaBB*
= *bbAAAAAAbb[Aba]aaaBcbaBAccBCaBBBBBBaaaaaaBB*
= *bbAAAAAAbbb[Aba]aaBcbaBAccBCaBBBBBBaaaaaaBB*
= *bbAAAAAAbbbb[Aba]aBcbaBAccBCaBBBBBBaaaaaaBB*
= *bbAAAAAAbbbbb[Aba]BcbaBAccBCaBBBBBBaaaaaaBB*
= *bbAAAAAAbbbbbbAB[Bcb]aBcbaBAccBCaBBBBBBaaaaaaBB*
= *bbAAAAAAbbbbbbA[Bcb]CaBcbaBAccBCaBBBBBBaaaaaaBB*
= *bbAAAAAAbbbbbbAcbCC[aA]ccBcbaBABBBBBBaaaaaaBB*
= *bbAAAAAAbbbbbbAcb[CCcc]BCaBBBBBBaaaaaaBB*
= *bbAAAAAAbbbbbbA[cbBC]aBBBBBBaaaaaaBB*
= *bbAAAAAAbbbbbb[Aa]BBBBBBaaaaaaBB*

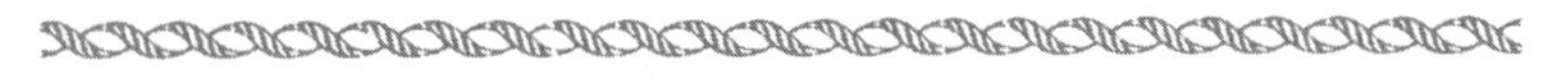

$= bbAAAAAA[bbbbbbBBBBBB]aaaaaaBB$
$= bb[AAAAAAaaaaaa]BB = [bbBB] = e$

For comparison, the reader can draw the given braid and try to unravel it geometrically—although doing so may instill an inferiority complex with regard to my notebook computer, which did the job in less than a tenth of a second.

Can Braids Be Used to Classify Knots?

Alexander's theorem affirms that all knots are closed braids, and we have just seen that braids can be classified. Can the classification of knots be deduced from these two facts? Several mathematicians, and not the least of them,[4] have probably nourished this hope (I know some who are still hoping). The attempt made for a rich history, brimming with new developments, that began during the 1930s and perhaps has not yet ended. But this chapter has gone on too long, and I will end it by referring the amateur math lover of nice stories to an article by Dehornoy (1997).

3

PLANAR DIAGRAMS OF KNOTS

(Reidemeister · 1928)

During the 1920s, the German mathematician Kurt Reidemeister, future author of the first book about the mathematics of knots, the famous *Knottentheorie*, began to study knots in depth. How could they be classified? The problem of systematizing the possible positions of a curve in space was to prove devilishly difficult.

The analytic approach (defining a knot using equations) did not help, nor did the combinatorial approach (defining a knot as a closed polygonal curve based on the coordinates of its successive vertices). In both cases, the information did not enable one to see the knot or to manipulate it. In practice, seeing a knot means drawing it, that is, projecting it onto a suitably chosen plane to obtain what is called a *knot diagram.* Manipulating the string that determines the position of the knot causes its diagram to undergo continuous modifications, which enables one to follow the evolution of its positions in space. But can the process be inverted? Can the projection be continuously modified in such a way as to obtain all the possible positions of the string in space? That is the question Reidemeister asked himself.

Here is his answer: Just perform a finite number of operations on the diagram, similar to those shown in Figure 3.1, while doing trivial planar manipulations (that is, while continuously changing the diagram of the knot in the plane without altering the number and relative disposition of the crossing points). The three operations shown in the figure are now called *Reidemeister moves;* they are denoted by the symbols Ω_1, Ω_2, Ω_3.The moves that they allow are the following:

- Ω_1: appearance (disappearance) of a little loop;
- Ω_2: appearance (disappearance) of twin crossings;
- Ω_3: passing a third strand over a crossing.

The following figure, which represents an unknotting process, shows how Reidemeister moves are involved in representing the manipulation of a knot. The process begins with the uncrossing (disappearance) of twin crossings at the top (shaded area of Figure 3.2a), followed by the passage over a crossing (b), the disappearance of a pair

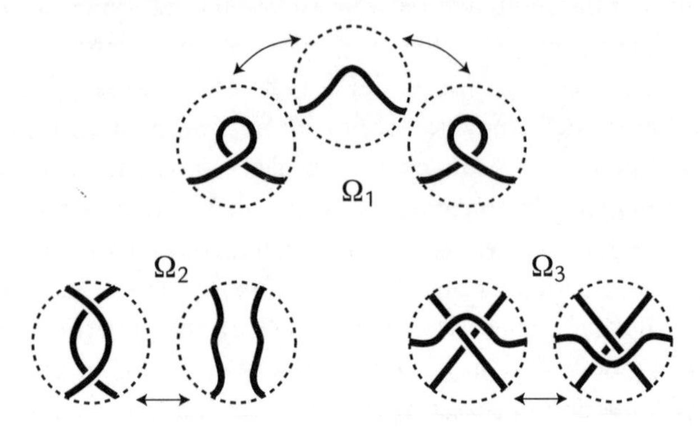

Figure 3.1. Reidemeister moves.

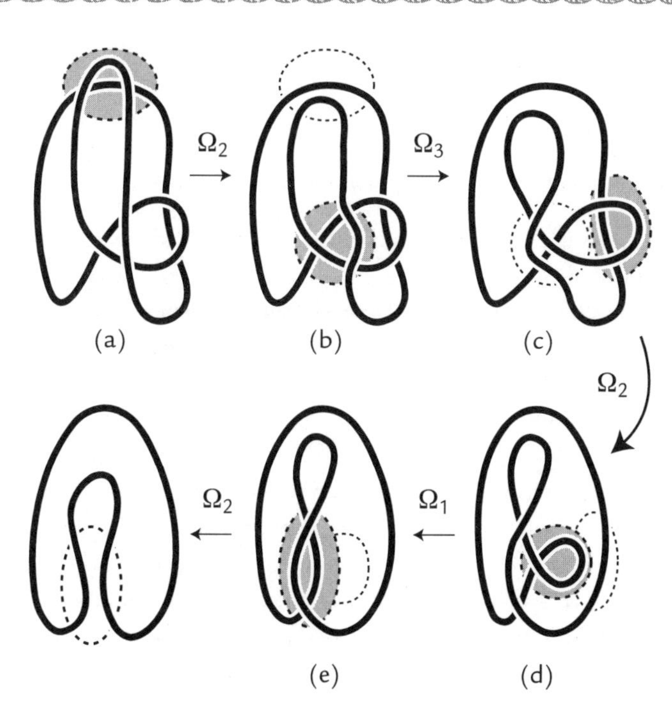

Figure 3.2. Unknotting using Reidemeister moves.

of twin crossings (c), the disappearance of a small loop (d), and, finally, the disappearance of a pair of twin crossings (e). The reader will have noticed that between the actual Reidemeister moves, the diagram of the knot undergoes trivial planar manipulations in preparation for these movements (these trivial manipulations change neither the number nor the distribution of the crossings).

To understand where Reidemeister moves come from, and why they

suffice, we will have to spend a little time talking about knot projections.

Generic Projection and Catastrophic Projection

In introducing the diagrams of knots, I mentioned above that they are projections in a "suitably chosen" plane. What does it mean to say that? A mathematician would answer that the plane must be selected so that the projection is *generic,* but this specification is utterly useless if you do not know the term, which is in fact a basic mathematical concept[1]—intuitively clear but difficult to formulate in the general case.

A generic projection is a projection without catastrophes, singularities, or degenerations that can be avoided, or gotten rid of by making tiny changes to the projected object. Let us be specific about what all these synonyms mean (catastrophe, singularity, degeneration) in the case of a knot represented by a closed polygonal curve. By definition (of a generic projection of a knot), we assume that

(1) two (or more) vertices cannot be projected to the same point;
(2) a vertex (or several vertices) cannot fall on the projection of an edge or line segment (to which they do not belong);
(3) three (or more) points cannot be projected to the same point.

The existence of a generic projection for any knot is obvious and can be demonstrated easily:[2] the "forbidden catastrophes" (1), (2), (3) must be removed by slightly displacing one of the knot's vertices. Note that these three catastrophes differ from the *crossing catastrophe,* which occurs when two points inside two distinct edges project onto a

single point: this is inevitable, inasmuch as any little change made to the knot moves the position of the crossing on the diagram somewhat, but cannot totally eliminate it.

Catastrophes (2) and (3) are, in some sense, the catastrophes that occur most often: among all the projections, they do represent exceptional events, but they are, so to speak, the most common among exceptional events. There are, of course, rarer catastrophes—also forbidden because they are special cases of catastrophes (1), (2), and (3). For example, 17 points, including 5 vertices, can project to a single vertex; 7 edges (perpendicular to the plane of projection) can degenerate to a single point; and so on.

Reidemeister moves correspond precisely to the most common forbidden catastrophes, as shown in Figure 3.3. Thus, in the upper part of Figure 3.3a, we see a type (2) catastrophe in which the vertex *A* of edge *AB* (which moves) projects momentarily (the middle frame) to a point inside the projection of edge *BC*. This position corresponds to the disappearance of a little loop on the projection (the knot diagram); that is, it is the move Ω_1. To the right of Figure 3.3a, this move is symbolically represented in keeping with the style of Figure 3.1. The shrewd reader, examining the next panels (b and c) of Figure 3.3, will see how a type (2) catastrophe can give rise to move Ω_2, and a type (3) catastrophe to move Ω_3.

The Sufficiency of Reidemeister Moves

Now that we know the origin of Reidemeister moves, we are in a position to discuss their principal application (often called Reidemeister's theorem or lemma):

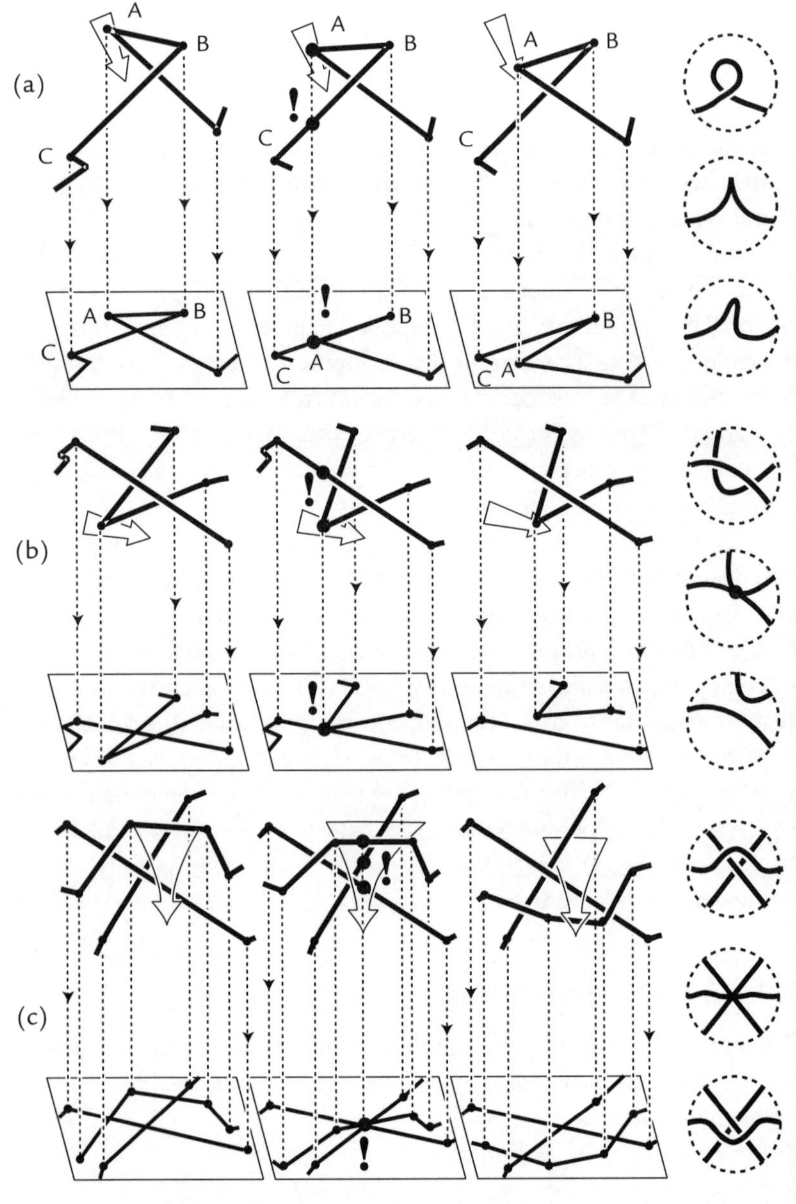

Figure 3.3. Catastrophes and Reidemeister moves.

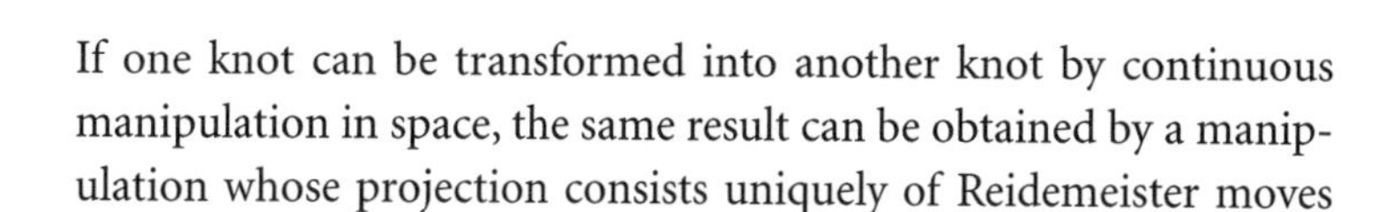

> If one knot can be transformed into another knot by continuous manipulation in space, the same result can be obtained by a manipulation whose projection consists uniquely of Reidemeister moves and trivial manipulations of the diagram in the plane.

This means that it is possible to examine all the spatial manipulations of knots by trivially manipulating their diagrams in the plane and applying Reidemeister moves from time to time. In this way, Reidemeister reduced the three-dimensional and rather abstract problem of knot equivalence to a two-dimensional problem, and a more concrete one, to boot.

Before discussing what the Reidemeister theorem contributes to the study of knots (in particular, to their classification), let me say a few words about its proof. Unfortunately (or fortunately, depending on your point of view), the proofs I know are not simple enough to be included in this book. For readers more at home with mathematics, I will simply say that all it takes is to analyze in detail a single basic triangular movement (see the beginning of Chapter 1) and to perturb it so that it is in a "general position"; in that case, only catastrophes (2) and (3) can occur and, as we have seen, these correspond exactly to Reidemeister moves.

Does the Reidemeister Theorem Classify Knots?

Let us put ourselves in Reidemester's shoes and imagine that we are elated at having demonstrated his theorem. Caught up in the excitement, we set out to classify knots—in other words, to devise an algorithm that will determine whether two knots (as sketched) are equivalent.

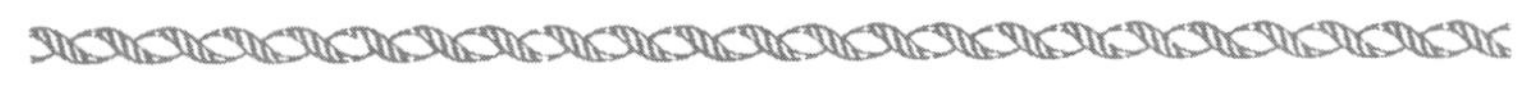

Take the first knot and compare it with the second. If the number of crossings and their relative positions are the same, the knots are equivalent, and we have succeeded. If not, (randomly) apply a Reidemeister move to the first knot and again compare the result with the second. If they are identical, we have won again. If not, we will have to remember the changes to the first knot and apply to it another Reidemester move, then compare again, and so on. If several moves applied to the first knot don't work, we go back to the modified knot, apply yet another move to it, compare, and so on. If the two knots are equivalent, sooner or later a set of Reidemeister moves leading from the first knot to the second will be specified.

The algorithm described above is easy to implement on a computer, even a very little one. Thus, the humble "notebook" on which I am writing this text has, among other things, software that can untie knots (by comparing them with the unknot, as in the preceding paragraph).[3] Does that solve the problem of classifying knots? Of course not, and for a good reason: applied to two nonequivalent knots, the algorithm described above never terminates; it continues on and on without ever stopping. And the user faces a dilemma: if the software does not give an answer after, say, a full day's work, is it because the knot is nontrivial or because the computer needs more time to find the sequence of operations that will lead to an unraveling?

Must we kiss Reidemeister's idea good-bye? Not yet, because there is another possibility that could salvage it. I am sure Reidemeister must have had this thought at one time or another, and perhaps it evoked in him that extraordinary feeling that researchers sometimes feel—the feeling of being on the verge of finding out, of understanding. (I note parenthetically that even for the best of us, this feeling is often followed by despair when the idea turns out to be inadequate or illusory.)

This other possibility is very simple: moves Ω_1, Ω_2, Ω_3 may diminish the number of crossings (disappearance of a small loop or of twin crossings), increase the number (appearance of a loop or twin crossings), or, finally, not change the number (Ω_3 simply passes a strand over a crossing). To create an algorithm for unraveling (that is, simplifying the knot), we just need the move Ω_3 and allow the moves Ω_1 and Ω_2 only when they decrease the number of crossings. If the process is restricted this way, the number of crossings diminishes, and the algorithm (thus perfected) will always terminate: either there are no more crossings (and the knot was trivial), or none of the permissible moves is applicable[4] (and the knot is nontrivial).

Unfortunately, this argument (however convincing) is erroneous. In reality one cannot always unravel a knot by simplifying it (by diminishing the number of crossings) at each step of the unraveling: sometimes, it is necessary at first to complicate things to make them simpler. An example of a trivial, nonsimplifiable knot (that can be unraveled only by first increasing the number of crossings) appears in Figure 3.4.

The hope of obtaining a simple and effective method of classifying knots using Reidemeister's theorem was too optimistic. The world is

Figure 3.4. A trivial, nonsimplifiable knot.

Figure 3.5. Wolfgang Haken's "Gordian knot."

made that way: to unravel a knotted situation, often one should begin by tangling it even further, only to unravel it better.

Since Reidemeister's theorem was discarded, devising trivial knots that are hard to unravel has admittedly become an important exercise in research on unknotting algorithms. A particularly barbaric example of such a knot (very hard to unravel—try it!) is shown in Figure 3.5, for which we can thank Wolfgang Haken. Moreover, it was Haken who finally solved the problem of unknotting (Haken, 1961), but his algo-

rithm (too complicated to be put on a computer) is based on a very different class of ideas.

What Remains of Reidemeister's Theorem?

We should not take the collapse of naïve expectations to mean that the application of Reidemeister's theorem is limited to an algorithm that does not work, or that this theorem is simply yet another example of a spectacular failure and dashed hopes.

The theorem occupied a central place in subsequent developments, especially in the study of knot invariants by Vaughan Jones, Louis Kauffman, and their followers (Chapter 6). In order that a knot diagram's function proposed in the guise of a new invariant be indeed invariant, the function must never change during the process of manipulating knots. According to Reidemeister's theorem, proving this requires only verification that the function does not change when the moves Ω_1, Ω_2, or Ω_3 are performed; well, these particular moves are very simple, so verification is in general very easy.

But that is not all. The failure of the unknotting algorithm described above is relative. Of course, from a theoretical point of view, it is not a real unknotting algorithm (since it can "continue indefinitely" without giving an answer). From a practical viewpoint, however, this algorithm and its recent modifications may be used as a relatively efficient tool that often enables a (strong enough) computer to unravel knots that cannot be undone "by hand"[5] . . . unless we use the algorithm that Alexander of Macedonia applied with so much success to the Gordian knot: cut it!

4

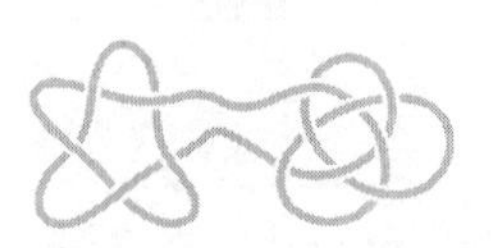

THE ARITHMETIC OF KNOTS

(Schubert · 1949)

Arithmetic—of knots? Absolutely. Why should the natural numbers—1, 2, 3, 4, 5, . . .—be the only objects that can be multiplied and decomposed into prime factors? The same features describe other mathematical entities, knots in particular. Moreover, knots possess an arithmetic very similar to that of the natural numbers, with commutative multiplication (called *composition*) and a theorem asserting the uniqueness of decomposition into "prime knots." The demonstration of this fundamental principle, foreshadowed by many researchers, proved difficult (as in fact did the corresponding principle for numbers) and was not achieved until 1949 (by the German mathematician Horst Schubert).

Thus, just as every whole number (say 84) factors out in a unique way ($84 = 2 \times 2 \times 3 \times 7$), every knot (for example, the one drawn at the left in Figure 4.1) is the (unique) composition of prime knots, as can be seen at the right in the same figure (it is the composition of two trefoil knots and one Turk's head knot). The reader will have grasped that the "composition" of knots consists more or less in setting them end to end (as is done for braids, anyone who has read Chapter 2 will say). To explain this operation, we are going to put knots[1] in boxes: in

Figure 4.1. Decomposing a knot into prime factors.

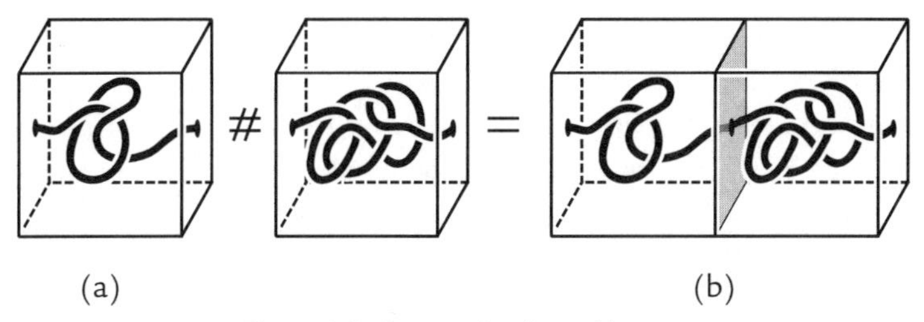

Figure 4.2. Composing boxed knots.

this way, each knot appears as a knotted string inside a cube, with the two ends stuck to two opposite faces of the cube (Figure 4.2a). (I will leave to the reader already corrupted by the study of mathematics the task of transforming this intuitive description into a rigorous mathematical definition.) Making a closed-curve knot from a boxed knot is easy (just join the two ends with a string outside the box), and vice versa. Once all the knots have been boxed, defining their composition is even easier: simply juxtapose the boxes and dissolve the double wall that separates them[2] (Figure 4.2b).

Our immediate goal is to study the main properties of the composition of knots. The first is associativity, which tells us that:

$$(a \# b) \# c = a \# (b \# c)$$

where the symbol # denotes the composition of the knots. This equation means that composing first the two knots *a* and *b* and then composing the knot obtained with the third knot *c* gives the same result as composing first the two knots *b* and *c*, then composing the knot obtained with *a*. This assertion is obvious, for it means, roughly speaking, juxtaposing the three knots (in both cases) and then eliminating the two walls (in a different order, of course, but the result is the same).

The following property is the existence of the *trivial knot* or *unknot*, indicated by the number 1, which (like the number 1) does not change the knot with which one composes it (just as the number 1 does not change the number it multiplies):

$$a \# 1 = a = 1 \# a$$

In its "boxed version," the unknot can be represented as a horizontal rectilinear thread in its cube. Juxtaposing such a box with that of any other knot obviously does not change the type of the knot.

The following property, being trickier, merits a subheading of its own.

Commutativity of the Composition of Knots

Like the multiplication of numbers, the composition of two knots is commutative (the result does not depend on the order of the factors):

$$a \# b = b \# a$$

This relationship is not at all obvious, but I am sure that its demonstration, shown schematically in Figure 4.3, will please the reader.

What is going on in this figure? First, pulling on the ends of the string that forms the first knot results in a small, tight knot (Figure 4.3b). Next, the little knot is slid along its own string, then along the string that forms the second knot (c). Still sliding along the string, the little knot traverses the big one and ends up to its right (d). Finally, the second knot is sent into the first box, and the little knot is blown up to original size. Voilà—there you have it (f)!

It may be hard for the reader to understand how a knot can "slide along a string." The simplest way is to take a good length of string (a shoelace will do) and execute the maneuver. Certain organisms are able to perform the process on themselves, which gives us the opportunity for another biological digression.

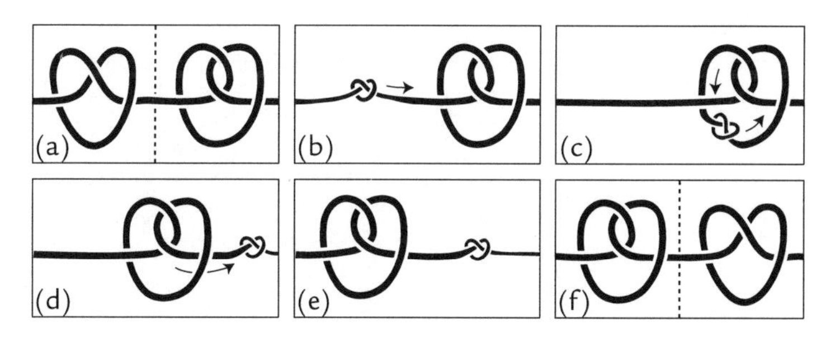

Figure 4.3. Composition doesn't depend on the order of the knots.

Digression: The Sliding Knot Fish

The strange fish in question is called a *myxine* (or, more commonly, slime eel). The myxine inhabits the ocean bottom in temperate latitudes. It has a supple backbone and secretes a very thick acidic saliva that it uses to coat its body when a predator tries to grab it. It quickly forms a knot with its tail and slides the knot along its body, thus spreading the saliva (which it secretes simultaneously) along its length (Figure 4.4a).

If you grab a myxine with your hand, it will slip through your fingers—not only because it is coated with oily saliva, but also be-

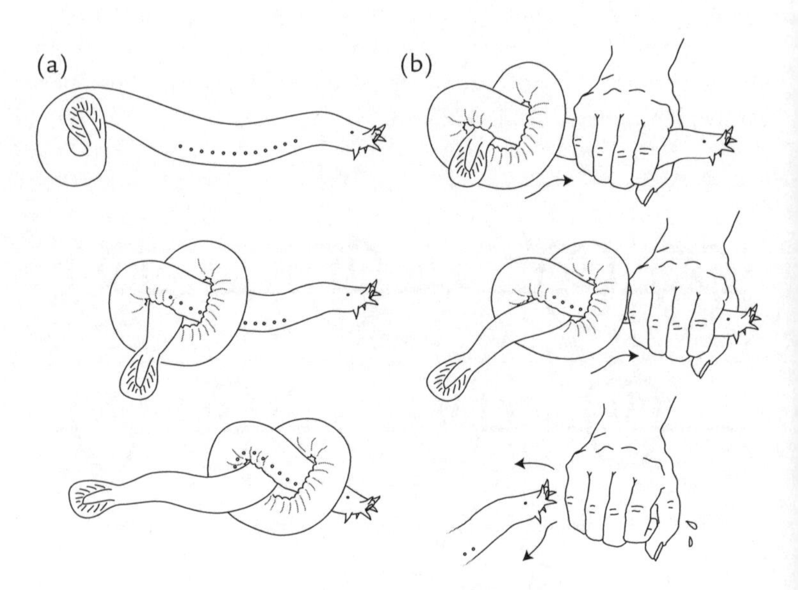

Figure 4.4. The myxine's knot.

cause of its knot, which it advances by pushing forcefully against your fist while its head moves toward the rear and escapes your grasp (Figure 4.4b). The traction it gets from moving its knotted body enables the myxine to carry out other vital tasks, especially feeding, which it achieves by suction (the myxine is a necrophage that leaves behind only the skin and bones of its prey).

Finally, once the danger has passed, the myxine gets rid of its layer of saliva (otherwise it would smother in its slimy cocoon) by the same sliding motion of the knot from the tail to the head. (For more details about this unusual animal, see Jensen, 1966.)

Note that the slime eel's knot is a trefoil (that is, the simplest nontrivial knot) and, in general, the left trefoil. To my knowledge, the myxine does not know how to make any other knots, but one could easily imagine a longer species of eel, with an even more flexible backbone, that would be capable of executing the same maneuver with more complicated knots.

But let us leave aside these biological knots and come back to their (admittedly more appetizing) mathematical models.

Can One Knot Cancel Out Another?

Having defined the composition of knots, one might ask whether inverse knots exist. That is, for a given knot, can one find another knot that, composed with the first, gives the unknot? In more geometrical language: If there is a knot at the end of a string, can a knot be made at the other end in such a way that the two knots cancel each other out when one pulls on the two ends of the string?

The analogous response to the question for natural numbers is no: for every natural number $n > 1$, there is no natural number m such

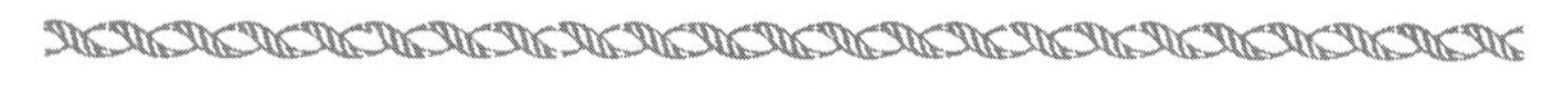

that $n \cdot m = 1$. (Of course, one can take $m = 1/n$, but then m would be a fraction, not a natural number.)

We will see that it is exactly the same with knots: no trivial knot possesses an inverse knot. This assertion is far from obvious. Indeed, it appears false from the start: Why can't one make a "symmetrical" knot at the other end of the string that would cancel out the first one?[3] Why? Before reading the explanation below, try some experiments with a string, beginning with the trefoil knot. Perhaps the ensuing failures will provide some clues for reflection.

Let us reason by contradiction: Let a and b be knots (nontrivial, that is, $a \neq 1$, $b \neq 1$) such that $a \# b = 1$. Consider the infinite composition:

$$C = a \# b \# a \# b \# a \# b \# a \# b \# a \# b \# a \# b \# \ldots$$

On the one hand, this composition is equal to the unknot, because we can write:

$$C = (a \# b) \# (a \# b) \# (a \# b) \# \ldots = 1 \# 1 \# 1 \# \ldots = 1$$

But by arranging the parentheses differently, we obtain:

$$C = a \# (b \# a) \# (b \# a) \# \ldots = a \# (a \# b) \# (a \# b) \# \ldots$$
$$= a \# (1 \# 1 \# \ldots) = a \# 1 = a$$

Thus we deduce that $a = 1$, which contradicts the assumption $a \neq 1$ and "shows" that there are no inverse knots.

The quotation marks in the preceding sentence indicate that in fact the "proof" is doubtful (that is the least one can say). In fact, coming back to whole numbers, one can "prove" in the same way that $1 = 0$:

just consider the infinite sum $1 - 1 + 1 - 1 + 1 - 1 + 1 - 1 + \ldots$ and place parentheses in two ways, precisely as above. The error in reasoning is the same in both cases: one cannot manipulate *infinite* sums or compositions (they must be defined beforehand) as one manipulates finite sums or compositions.

In the case of knots, however, a slight modification of the argument makes it totally rigorous. Just place copies of the knots a and b one by one in an infinite series of boxes that become smaller and smaller and converge to a point—which, by the way, rigorously defines the infinite composition C (Figure 4.5)—and replace the questionable arithmetic manipulations by correctly defined topological manipulations. I will skip the technical details.[4] The reader will have to take my word that this clever proof using infinite compositions is more than a brilliant sophism (like that of Achilles and the tortoise); indeed, it is the basis of a rigorous mathematical argument.

Finally, to end this section, let me add that the author of this clever argument is not a specialist in knot theory but the German philosopher and politician Wilhelm Leibniz, a great mathematician when he put his mind to it (inventor, independently of Newton, of the differential and integral calculus). Leibniz discovered this argument in an entirely different context, since knot theory did not exist at the time, in order to prove (correctly) a theorem concerning a classical object in the differential calculus—conditionally convergent series.

Prime Knots

We have just seen that there are no inverse knots, just as there are no inverse natural numbers (which means, in other words, that the num-

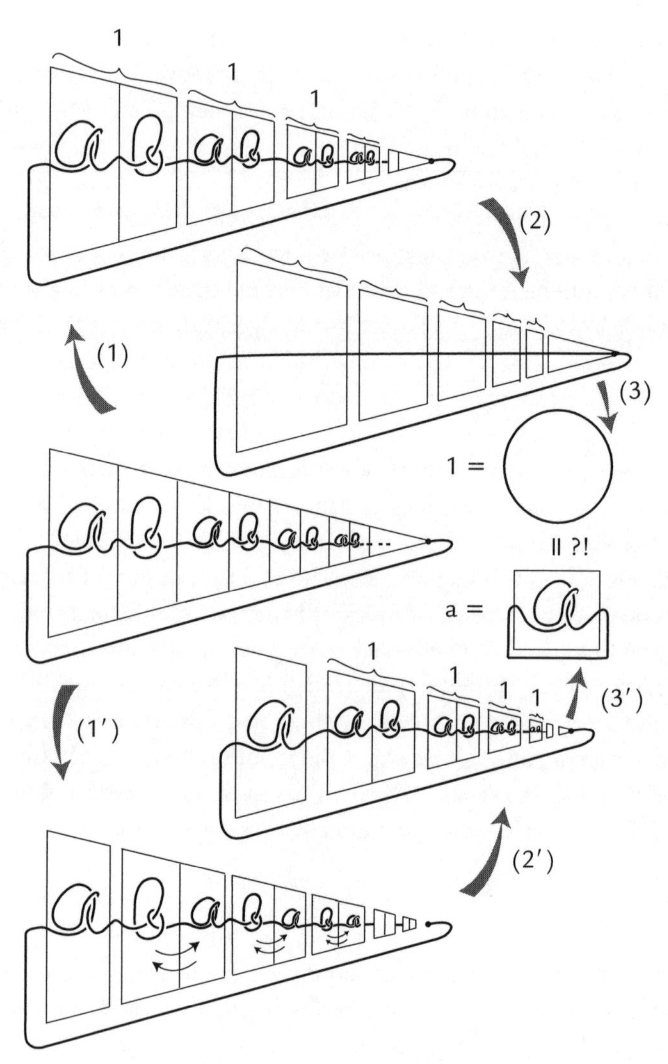

Figure 4.5. One knot cannot cancel out another.

ber 1 has no divisor other than itself). This property (of having no divisors other than itself and 1) is in fact the definition of prime numbers, in principle taught in the earliest grades at school. This simple definition gives the following mysterious set:

2, 3, 5, 7, 11, 13, 17, 19, 23, 29, 31, 37, 41, 43, 47, 53, . . .

The set of prime numbers has continued to mystify mathematicians since the beginning of their profession. But what is the situation for knots? Are there prime knots, knots that cannot be represented as the composition of two other nontrivial knots? The answer is yes: the trefoil, the figure eight knot, and the two alternating knots with five crossings (Figure 4.6) are prime, whereas the square knot (called a double knot by boy scouts) and the granny knot are composite knots.

How can we establish that prime knots exist? How does one prove, for example, that the trefoil is indeed a prime knot? The notion that comes immediately to mind is to use the minimal number of knot crossings: if the trefoil (which has three crossings) is the composition

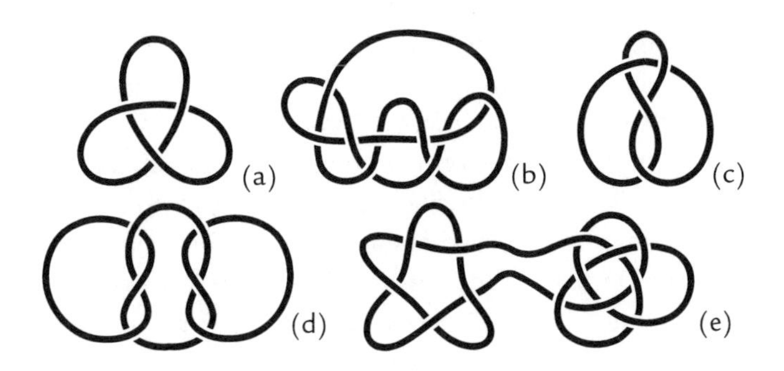

Figure 4.6. Prime knots and composite knots.

of two other nontrivial knots, those would have at least three crossings each (because knots with two crossings or fewer are trivial). Two times 3 makes 6, 6 is greater than 3, and so there is a contradiction: QED. Unfortunately, this argument is insufficient because we do not know whether the minimal number of crossings of a composite knot is equal to the sum of the minimal number of crossings of the two factors. This assertion is actually true, but its proof is too difficult to include in this book.

Thus, every knot decomposes into prime knots. Now, the decomposition of each natural number into prime factors is unique. How does that work for knots?

Unique Factorization into Prime Knots

Here, too, the parallelism with natural numbers is complete: *every knot decomposes uniquely into prime knots.* Obtaining the proof of this marvelous theorem was the ambition of many researchers. It was Horst Schubert who did it at the end of the 1940s, though his proof, at once profound and very technical, is beyond the scope of this book.

Schubert's theorem, which is associated with other properties common to knots and to natural numbers, brings us to the obvious idea of numbering knots in such a way as to conserve the decomposition into prime factors. Such a numbering would associate a prime number to each prime knot, a composite number to each composite knot, such that the prime factors of the number are the prime factors of the knot. Alas! such numbering may indeed exist in principle, but there is no natural algorithm that can produce it.

The reason is that, in contrast to numbers, two knots cannot be

added; they can only be multiplied (composed). Each positive whole number can be obtained by adding an appropriate quantity of ones ("trivial numbers for multiplication"); thus $5 = 1 + 1 + 1 + 1 + 1$. On the other hand, it is not possible to obtain all knots by "adding copies of the trivial knot," since this operation of addition does not exist.[5]

Another reason there is no natural numbering of knots is the lack of order in the set of knots. Natural numbers possess a natural total order (1, 2, 3, 4, 5); this order does not exist (or has not been discovered!) for knots. Of course, it is usual to order knots by the number of minimal crossings of their diagrams, but this order is not linear: which of the two knots with five crossings (see the table of knots in Figure 1.6) is the "smallest"?

So the arithmetic of knots has not helped us to classify them. But there is scant reason to talk of failure here: Schubert's theorem does not need any applications; it is mathematical art for art's sake, and of the most exalted kind.

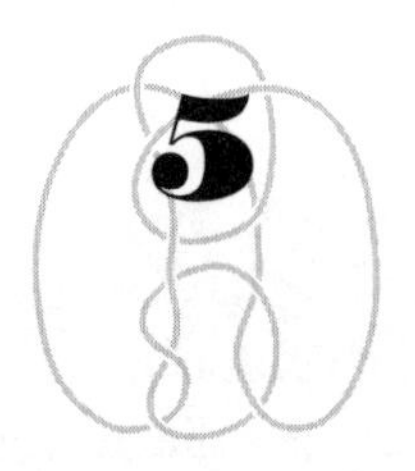

SURGERY AND INVARIANTS

(Conway · 1973)

In 1973, the English mathematician John Conway discovered the fundamental role played in knot theory by two very simple "surgical operations," two ways of modifying a knot near one of the crossings of its strands.

The first, which we will call the *flip,* consists in transforming the chosen crossing (on the planar representation of the knot) into the opposite crossing, that is, running the upper strand under the lower strand (Figure 5.1); with a string, the flip can be performed by cutting the upper strand, then reattaching it under the other strand. Of course, the flip can change the type of the knot; for example, "flipping" one of the crossings of a trefoil knot produces a trivial knot (the trefoil unknots—try to draw it, and you'll see).

Conway's second little surgical operation, called *smoothing,* consists in eliminating the crossing by interchanging the strands (Figure 5.2a); with a string, it is done by cutting the two strands at the point of crossing and reattaching them "the wrong way" (Figure 5.2b). Note that when the strands are not oriented, there are two ways of reattaching the four ends two by two (b or c), but the orientation of the knot

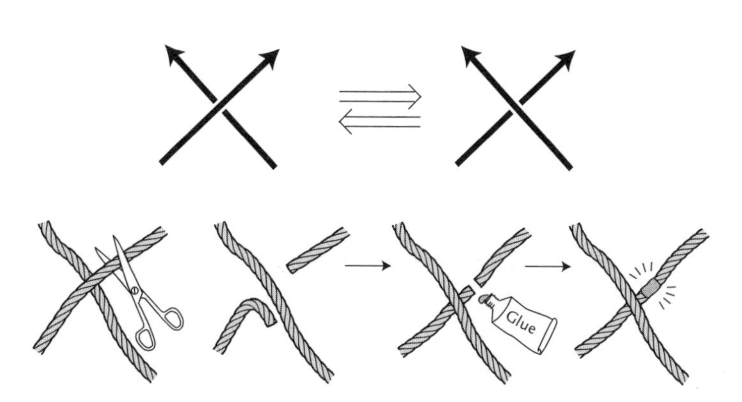

Figure 5.1. The flip (the upper strand becomes the lower strand).

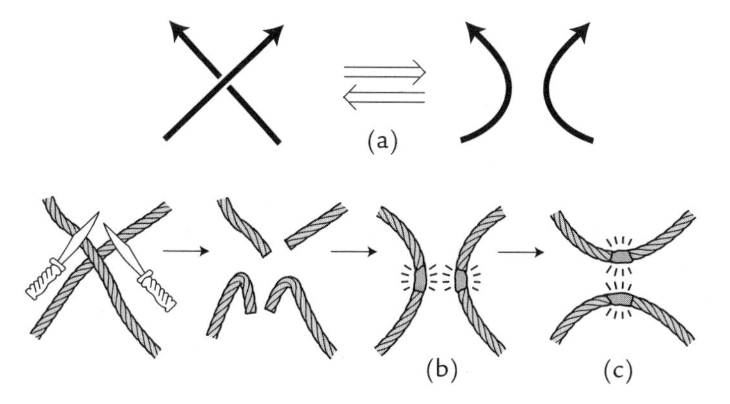

Figure 5.2. Smoothing (the strands are reattached the wrong way).

leaves us with no choice as to which pairs of ends should be reattached (this is dictated by the arrows, as in Figure 5.2a).

The flip and the smoothing operation were known and often used by topologists well before Conway; in particular, J. W. H. Alexander used them to calculate the polynomials that bear his name (to which we will return later). Conway's contribution was to show that these two operations can serve as the basis for a fundamental definition of a knot invariant (the Conway polynomial), which we will take up a little further in this chapter.

Actually, the importance of Conway's operations extends far beyond the scope of knot theory. Flipping and smoothing play an essential role in life itself, as they have been routinely used by nature ever since biological creatures began to reproduce. The description of this role—which is quite extraordinary—merits at least a small digression.

Digression: Knotted Molecules, DNA, and Topoisomerases

Watson and Crick's seminal discovery of DNA, the molecule that carries the genetic code, presented biochemists with a series of topological puzzles, among other problems. We know that this long, twisted double helix is capable of duplicating itself, then separating into two identical molecules that—unlike the two constituents of the parent molecule—are not bound together but are free to move. How is that possible topologically?

Intricate studies have shown that enzymes called *topoisomerases* specialize in this task. More specifically, topoisomerases perform the three basic operations shown in Figure 5.3. The reader will immedi-

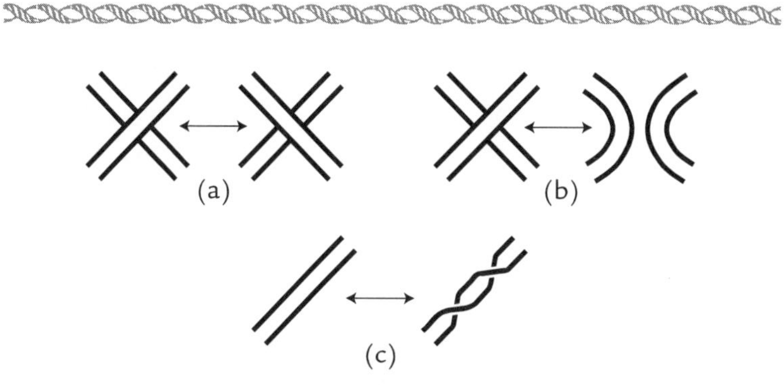

Figure 5.3. Topoisomerase operations on DNA.

ately recognize operations (a) and (b): they are the flip and Conway's smoothing! The third operation (Figure 5.3c), which is called the *twist,* is also known in topology; it refers to the mathematical theory of ribbons, which is currently very useful in theoretical physics.

Let us look more closely at how these strange enzymes affect long molecules, especially DNA molecules. First of all, note that the rearrangement of the strands occurs at the molecular level and is not visible: even the most powerful electron microscopes provide only indirect information.

Remember, too, that the DNA molecule appears in the form of a long double helix, each strand of which is constituted of subunits, the bases A, T, C, and G, whose arrangement on the strand codes the genetic properties of the organism (a little like the way the arrangement of the numbers 0, 1, 2, . . . , 9 on a line of text gives the decimal code of a number). Figure 5.4 is a schematic representation of a fragment of a DNA molecule.

It is well known that the double strands of DNA ordinarily have free

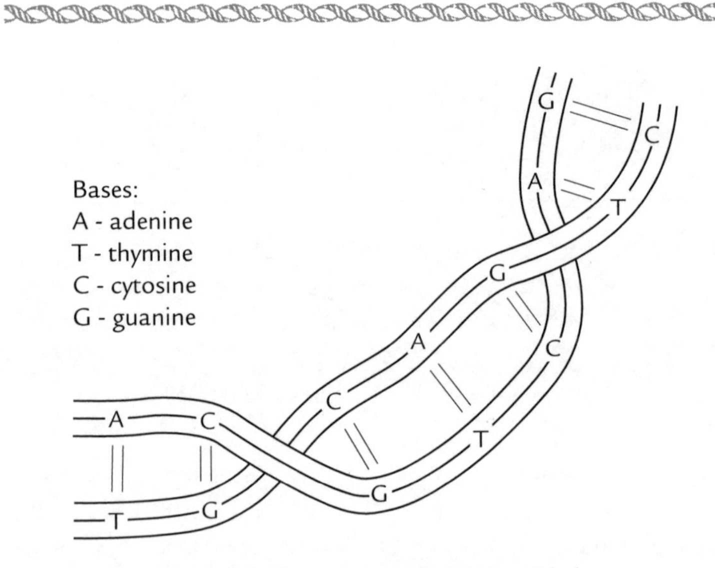

Figure 5.4. The structure of a DNA molecule.

ends, but it is not always so; there are also molecules with closed double strands (the twice-coiled snake biting its tail) and single-stranded molecules with both closed and open ends. These molecules take part in three classical genetic processes—*replication, transcription,* and *recombination;* moreover, the double-stranded molecules form supercoils or tangles (this transforms these drawn-out objects into compact ones). The topoisomerases play a crucial role in this whole process by carrying out the tasks of cutting, rearranging, and rejoining. First, the enzymes nick one of the strands, pass the second strand through the opening, then reattach the cut so that the strands change place (Conway's flip). Conversely, by means of two cuts and two repairs, the enzymes are able to reattach the two strands "the wrong way" (Conway's smoothing).

The precise mechanism of the cutting, rearranging, and rejoining operations is still not well understood. But it is known that there are different types of topoisomerases (they are not the same for single- and double-stranded DNA). And we have some idea, from the work of James Wang, how supercoiling (and the reverse operation) of a closed double-stranded DNA molecule works.

DNA supercoiling is analogous to what often happens to a telephone whose handset is connected with a long spiral cord. When the user twists the cord while returning the handset to the base, the connecting wire gets more and more tangled, eventually becoming a sort of compact ball. For the phone user, this result is rather annoying, since the coiling shortens the distance between the phone and the handset. In the case of DNA, supercoiling also transforms the long spiral into a compact ball, but here the result is useful, because transforming the long molecule (several decimeters) into a tiny ball makes it easy for the molecule to enter the nucleus of a cell, whose dimensions are measured in angstroms.[1]

In its normal state (not supercoiled), the DNA spiral makes a complete rotation with each series of 10.5 successive bases. In making twists (take another look at Figure 5.3c), the corresponding topoisomerase transforms the simple closed curve of DNA in the manner indicated in Figure 5.5. Note that, from the topological point of view, one of the results of the twist is to change the *winding number* of the two strands of DNA (this invariant, which we owe to Gauss, measures the number of times one of the strands wraps around the other). There

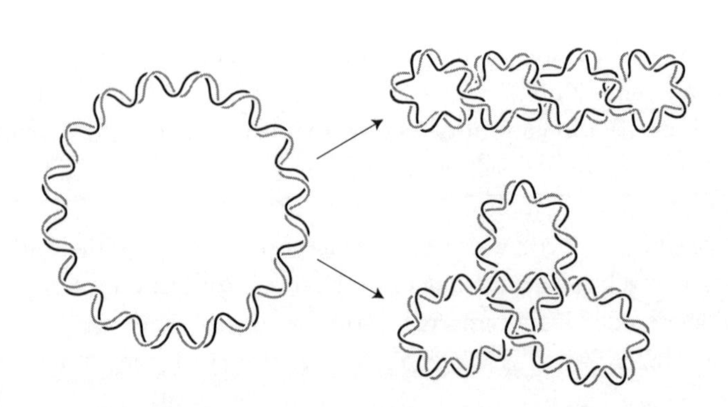

Figure 5.5. A supercoiled, double-stranded DNA molecule.

are many other topological phenomena that fascinate biologists, but our goal is not to give a detailed description of the findings. For a more thorough introduction, the reader is referred to Wang (1994).

Invariants in Knot Theory

Let us return to the mathematical theory of knots, to speak—at last—of invariants. How do they appear, and what role do they play in the theory?

Roughly, knot invariants serve above all to respond, when needed, in the negative to the most obvious question regarding knots, which we have called the *comparison problem:* Given two planar representations of knots, can we say whether they represent the same or different knots? For example, drawings (a) and (e) in Figure 5.6 represent the same knot—the trefoil. Indeed, this same figure shows how representation (a) can be converted to representation (e). On the other hand,

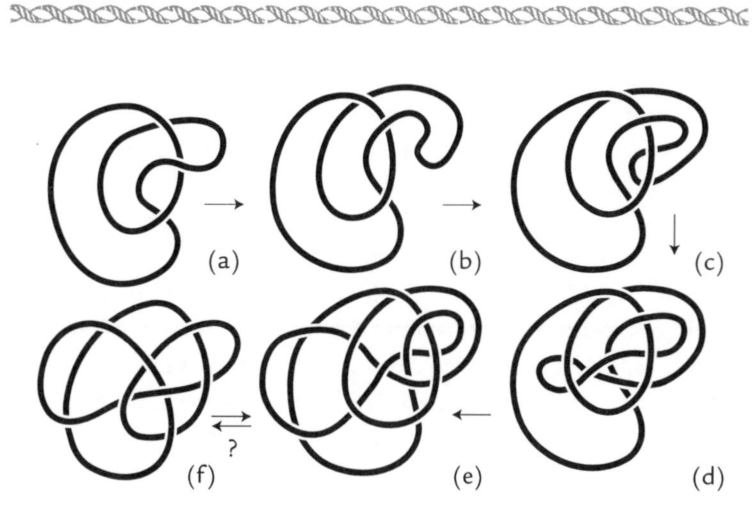

Figure 5.6. Six representations of the same knot?

any attempt to transform drawing (f) into a representation of a trefoil is doomed (try it!). But how do we prove it? The fact that we have not succeeded in turning one sketch into another does not prove anything; a more clever person, or a luckier one, might be able to do it easily.

Now suppose that we have at our disposal a knot invariant, that is, a way of associating, with each planar representation of a knot, a certain algebraic object (a number, a polynomial) so that this object never varies when the knot is manipulated, as in the first five sketches (a–e) in Figure 5.6. When two planar representations are given (for example, f and e in Figure 5.6), one can calculate their invariants. Differing values for the invariant prove that the two representations do not define the same knot (if they did, they would have the same invariant!).

For example, the calculation of Conway's polynomial (to be described in more detail later) of representations (a) and (f) in Figure 5.6 gives $x^2 + 1$ and $-x^4 + x - 1$, respectively; therefore, the diagrams in question clearly represent two different knots.

Before moving on to the study of Conway's invariant, let us try to find a numerical invariant for knots ourselves. The first idea that comes to mind is to assign to each knot diagram the number of its crossings. Unfortunately, this number is not an invariant: as one manipulates a knot in space, new crossings may appear on its planar projection, and others may disappear (for example, see Figure 5.6). Those who have read Chapter 3 may remember that the first and second Reidemeister moves change the number of crossings by adding ± 1 and ± 2, respectively.

But it is easy, using this idea, to try to find a genuine invariant of the knots: just consider the *minimal* $c(N)$ number of crossings of all the projections of knot N. This number (a nonnegative integer) is by definition an invariant (it does not depend on the specific projection given, since the definition involves all projections). Unfortunately, it is useless for comparing knots, because it would work only if we could calculate $c(N)$ from a given projection N. Since we do not presently have any algorithm for doing this calculation, we will move on to a more sophisticated but calculable invariant: Conway's.

Conway's Polynomial

To every planar representation N of an oriented knot, Conway associates a polynomial in x, noted $\nabla(N)$, which satisfies the three following rules:

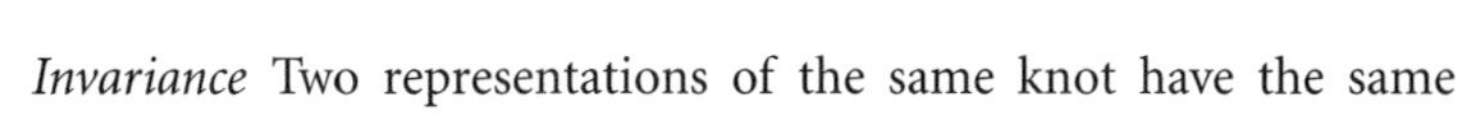

Invariance Two representations of the same knot have the same polynomial:

$$N \sim N' \quad \Rightarrow \quad \nabla(N) = \nabla(N') \qquad \text{(I)}$$

Normalization The polynomial of the unknot is equal to 1 (it is regarded as a "zero-order polynomial"):

$$\nabla(\bigcirc) = 1 \qquad \text{(II)}$$

Conway's skein relation Where the three planar representations N_+, N_-, and N_0 are identical outside the neighborhood of one crossing and have forms as indicated in the diagrams as follows:

N_+: $\qquad$ N_-: $\qquad$ N_0:

then

$$\nabla(N_+) - \nabla(N_-) = x\,\nabla(N_0) \qquad \text{(III)}$$

(In other words, N_0 and N_- are derived from N_+ by a smoothing and a flip, respectively.)

For example, when N_+ defines the trefoil knot, the Conway notation specifically gives:

$$-\nabla(\;) + \nabla(\;) = x\nabla(\;) \qquad (5.1)$$

The attentive reader will have noticed that in this case the diagram N_0 (on the right side of the equal sign) no longer describes a knot: it consists of two closed curves instead of one, and it is the diagram of a link (a family of curves in space that can knot separately as well as link together). But never mind that—Conway's polynomial is actually defined for all links (of which knots are just a particular case).

From now on, we will write Conway's skein relation (and other similar relations) in the following symbolic form:

$$\nabla(\ ) - \nabla(\ ) = x\nabla(\ )$$

which means that there are three identical links outside the circular areas (bounded by the dotted lines) that each contain one crossing. The second and third links are obtained by flipping and smoothing the first link in this neighborhood.

Examples of the Calculation of the Conway Polynomial

One of the advantages of the Conway invariant is the ease with which it can be calculated. Here are some examples.

Consider the link consisting of two separate circles. Then we have $\nabla(\bigcirc\bigcirc) = 0$. In fact:

$$x\nabla(\ ) \overset{(III)}{=} \nabla(\ ) - x\nabla(\ ) \overset{(II)}{=}$$
$$\overset{(II)}{=} \nabla(\bigcirc) - \nabla(\bigcirc) \overset{(I)}{=} 1 - 1 = 0$$

Now assume that the links consist of two linked circles, known as the *Hopf link*, $H =$  . Following Conway's notation,

$$\nabla(\ ) - \nabla(\ ) \overset{(III)}{=} x\nabla(\ )$$

and since $\nabla(\bigcirc\bigcirc) = 0$ and $\nabla(\bigcirc) \overset{(I)}{=} 1$, we deduce that $\nabla(H) = x$.

Finally, let us calculate Conway's polynomial for the trefoil T. For that, we return to equation 5.1; by virtue of rule I, the second term on the lefthand side is equal to $\nabla(\bigcirc)$, and by virtue of rule II, to 1; the term in the righthand side, according to the preceding calculation, is equal to $x \cdot x = x^2$. Thus one obtains $\nabla(T) = x^2 + 1$.

Expressed in this way, the calculation of Conway's polynomial for a knot (or link) appears to be a rather original mix of geometrical operations (namely, the flip and the smoothing) and classical algebraic operations (sums and products of polynomials). The reader with a taste for this type of thing will undoubtedly enjoy calculating $\nabla(P)$, where P is the knot represented in Figure 5.6f.

Discussion of Results

What can we deduce from these calculations? Many things. In particular, we now have formal proof that:

(1) The two curves of a Hopf link cannot be separated:

(2) A trefoil knot cannot be unknotted:

(3) The knot represented in Figure 5.6f is not a trefoil:

Of course, the reader who has not yet been ruined by mathematics will say that there is no point in having a formal proof of something so obvious as statements (1), (2), or (3). But one can argue that what we have here is a general method that also works in more complicated situations, where intuition proves of no avail.

For example, for the planar representations of Figure 5.7, my intuition of space (fairly well developed) does not necessarily tell me anything about the knot(s?) they represent. Yet my little notebook computer, which has software for calculating Conway's polynomial, trumpeted after a few seconds' reflection that $\nabla(A) = 1$ and $\nabla(B) = x^2 + 1$, which proves that the knots represented by A and B are different.

Thus we have in Conway's polynomial a powerful invariant that allows us to distinguish knots. Does it always work? In other words, does the equality of the polynomials of two knot representations imply that they are two representations of the same knot? Does one always have $\nabla(K_1) = \nabla(K_2) \Rightarrow K_1 = K_2$? Unfortunately, the answer is no: a calcula-

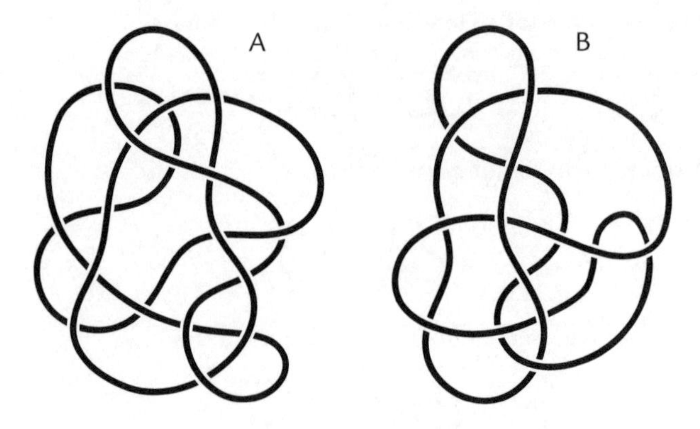

Figure 5.7. Two representations of the same knot?

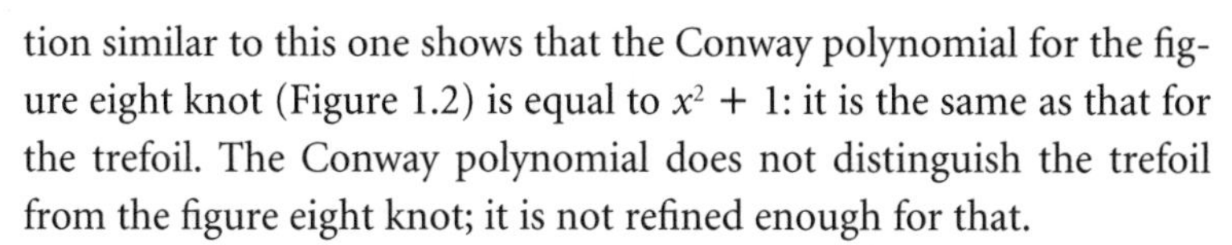

tion similar to this one shows that the Conway polynomial for the figure eight knot (Figure 1.2) is equal to $x^2 + 1$: it is the same as that for the trefoil. The Conway polynomial does not distinguish the trefoil from the figure eight knot; it is not refined enough for that.

But—the sceptical reader will counter—what tells us that the trefoil and the figure eight knot are not, in fact, the same knot? Good question. We will only be able to answer it conclusively when we have an invariant more sensitive than Conway's. One example is Jones's famous two-variable polynomial (the topic of the next chapter) or the Homfly polynomial, which can also be obtained using Conway's method and with which we will end this chapter.

The Homfly Polynomial

"Homfly" is not the name of the inventor of this polynomial: it is an acronym for the six(!) researchers who discovered the same polynomial at the same time and published their results simultaneously (in 1985) in the same journal. They are H = Hoste, O = Ocneanu, M = Millet, F = Freyd, L = Lickorish, and Y = Yetter.[2]

The simplest way to define the Homfly polynomial $P(x, y)$ (with two variables x and y) is to use Conway's axioms I, II, and III with P in the place of ∇ and with the following modification to the skein relation (axiom III):

$$xP\left(L_+\right) - yP\left(L_-\right) = P\left(L_0\right). \qquad \text{(III}'\text{)}$$

The reader who has grasped how to perform the little calculations of Conway polynomials of knots will perhaps enjoy redoing these calculations with the new skein relation III′ for the same knots and links. In particular, he will then see that the Homfly polynomials for the trefoil and the figure eight knot are not the same.

The Homfly polynomial is thus more sensitive than Conway's. But is it a complete invariant? Can it distinguish all nonisotopic knots? Unfortunately, the answer is no: Figure 5.8 shows two different knots that have the same Homfly polynomial.

That is why the search for a complete invariant continues in the following chapters.

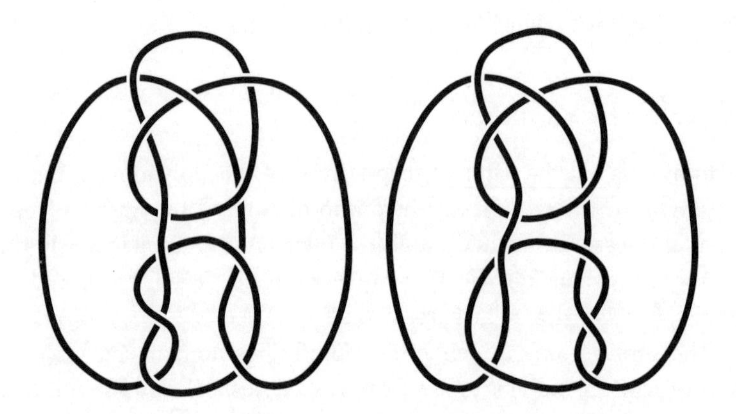

Figure 5.8. Two knots that have the same Homfly polynomial.

JONES'S POLYNOMIAL AND SPIN MODELS

(Kauffman · 1987)

The discovery by Vaughan Jones of the polynomial that bears his name reinvigorated the study of knot invariants. Of that there can be no doubt. But the significance of the famous polynomial extends far beyond the scope of knot theory: it is popular because of its connections to other branches of mathematics (the algebra of operators, braid theory) and especially to physics (statistical models and quantum groups).

It thus makes sense to devote this chapter—a key one—to Jones's theory. Unfortunately, as conceived by its author at the outset, this theory is far from elementary (see Stewart, 1989), and it exceeds, by a long shot, the assumed mathematical sophistication of the wide readership I hope to reach. But it so happens that another approach to Jones's polynomial, that of Louis Kauffman of the University of Chicago, has the double advantage of being very easy and of clearly showing the relationship of the polynomial with statistical physics. It is on this branch of physics that I shall base an explanation of the theory, and so I will begin with a few of its fundamental notions.

Statistical Models

For a good thirty years (and especially since the publication in 1982 of Roger Baxter's book on this subject), statistical models and in particular the famous Ising model have captured the attention of both mathematicians and physicists. What is it all about? It has to do with theoretical models of regular atomic structures that can adopt a variety of states, each state being determined by the distribution of spins on the atoms (a very simple example is shown in Figure 6.1). At any given instant, each atom (represented in the figure by a fat dot) is characterized by its interactions with its neighboring atoms (an interaction is represented by a line joining the two atom-dots) and its "internal state"—what physicists call its "spin," a parameter[1] that has a finite number of values (in the model here, two). The two spins in this model are *up* and *down* and are represented by arrows pointing up and down, respectively.

For the model to be wholly deterministic, we must specify its partition function. This is an expression of the form:

$$Z(P) = \sum_{s \in S} \exp \frac{-1}{kT} \sum_{(a_i a_j) \in A} \varepsilon[s(a_i), s(a_j)] \tag{6.1}$$

where the external sum is taken over the set S of all states, and the internal sum is taken over all edges (interactions), while $\varepsilon[s(a_i), s(a_j)]$ is the energy of the interaction between the atoms a_i and a_j (which actually depends only on their spins), T is the temperature, and k a coefficient called *Boltzmann's constant* (whose value depends on the choice of units).

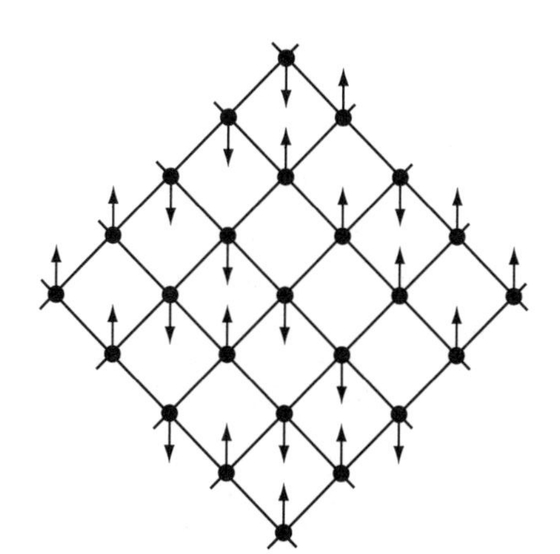

Figure 6.1. The spin model.

Using the partition function Z, we can calculate a given model's total energy and the probability that it is in a given state and, especially, study its phase transitions—for example, in the case of the Potts model of freezing water, its change from the liquid state (water) to the solid one (ice) and vice versa.[2]

I do not intend to delve further into the study of statistical models. The little bit I have just said about it will be enough for the reader to understand where Louis Kauffman got his funny idea of associating a statistical model with each knot.

Kauffman's Model

Take any (unoriented) knot, such as the one shown in Figure 6.2. Look closely at one of the knot's crossings; locally, each intersection divides the plane into two complementary angles, one of which is *type A* (or *up type*) and the other *type B* (or *down type*). The type-A angle is the one we see to our right when we start moving along the upper strand, and then (after we cross over the lower strand) to our left. (The direction chosen for crossing the intersection can be either of the two possibilities: the resulting type of angle does not depend on this choice—check that!) In Figure 6.2, A angles are shaded, and B angles are left blank.

Thus, for a given knot, we can choose at each intersection what might be called a Kauffman spin. In other words, we can associate with each crossing the word *up* or the word *down.* We say that such a choice (at all crossings) is a *state* of our knot. A knot with n intersections thus has 2^n possible states. To represent the knot in a specific state, we could have written *up* or *down* next to each intersection, but I prefer to draw a little stick inside the chosen angle (look again at Figure 6.2a, as well as Figure 6.2b).

This notation also has the advantage of clearly indicating the choice between one of two ways of smoothing a crossing (of an unoriented knot) by exchanging strands "following the little stick" (Figure 6.3b). We will need to make this choice right away. Let us denote by $S(K)$ the set of all the states of the knot K. To completely define the Kauffman model associated with the knot K, we need only define the corresponding partition function. It will be denoted by $\langle K \rangle$, called the *Kauffman bracket,* and defined by the equation:

$$\langle K \rangle = \sum_s a^{\alpha(s)-\beta(s)}(-a^2 - a^{-2})^{\gamma(s)-1} \qquad (6.2)$$

where the sum is calculated for all the possible 2^n states $s \in S(K)$ of the knot K, where $\alpha(s)$ and $\beta(s)$ denote the number of type-A and

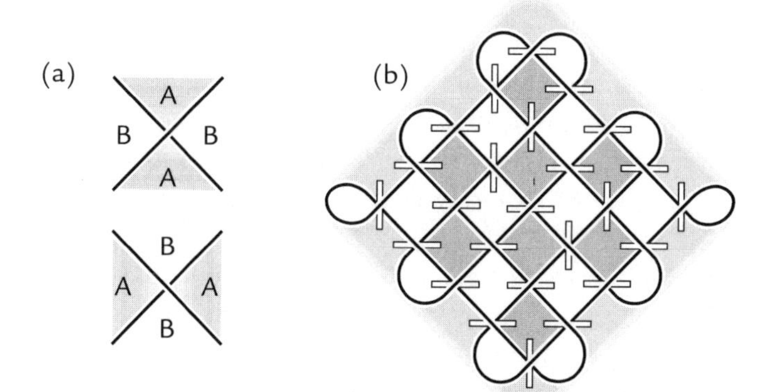

Figure 6.2. State of a knot, illustrating angles of type A and B.

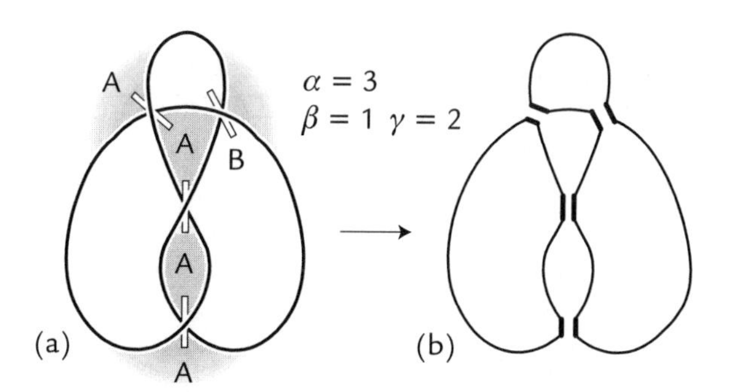

Figure 6.3. Smoothing a figure eight state.

type-B crossings, respectively, whereas $\gamma(s)$ denotes the number of closed curves obtained when all the knot crossings are smoothed following the little s-state sticks.

One might ask where Kauffman got this unusual formula (which is quite unlike its prototype in equation 6.1). Without going into details, I would simply say that he found it by backward reasoning (and trial and error), that is, by starting from the result he wanted to obtain.

Be that as it may, the application of this formula is very simple (though very laborious if the knot has many crossings). Figure 6.3 shows a possible state for the figure eight knot diagram (a) and the result of the corresponding smoothing (b); for this example, $\alpha(s) = 3$, $\beta(s) = 1$, and $\gamma(s) = 2$ (two closed curves appear after all the crossings are smoothed). Consequently,

$$a^{\alpha(s)-\beta(s)}(-a^2 - a^{-2})^{\gamma(s)-1} = a^{3-1}(-a^2 - a^{-2})^{2-1} = -a^4 - a^1$$

Obtaining the value of Kauffman's bracket for the figure eight knot diagram requires drawing all 16 possible states of the diagram ($16 = 2^4$) and summing the 16 results for the equation above (which describes just one of the 16 states). In this way, we get a polynomial[3] (in a), which is the value of the Kauffman bracket for the diagram of the given knot.

Note in addition that equation 6.2 is also valid for links of more than one component.

Before continuing with our study of the Kauffman bracket, let us pause a moment to compare the result we get with a classical model, such as the Potts model. Let us begin by comparing Figures 6.1 and 6.2. Aren't they awfully similar? Of course, the graphical similarity re-

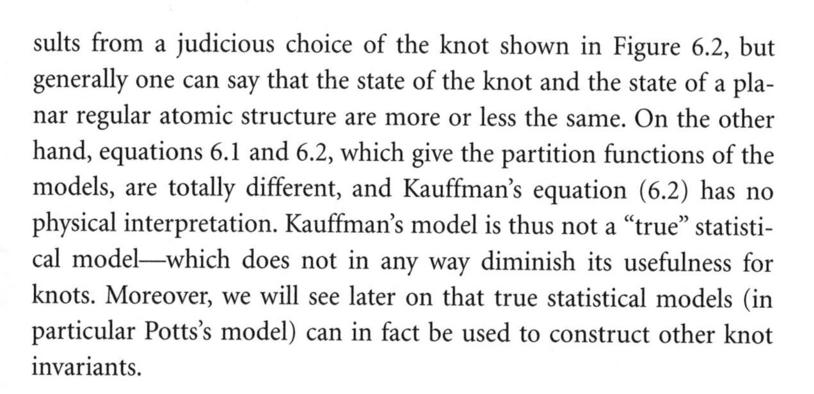

sults from a judicious choice of the knot shown in Figure 6.2, but generally one can say that the state of the knot and the state of a planar regular atomic structure are more or less the same. On the other hand, equations 6.1 and 6.2, which give the partition functions of the models, are totally different, and Kauffman's equation (6.2) has no physical interpretation. Kauffman's model is thus not a "true" statistical model—which does not in any way diminish its usefulness for knots. Moreover, we will see later on that true statistical models (in particular Potts's model) can in fact be used to construct other knot invariants.

Properties of the Kauffman Bracket

Our first goal is to indicate certain properties of this bracket to see how to deduce an invariant of knots from the bracket.

The three principal rules are the following:

$$\langle \times \rangle = a\langle \asymp \rangle + a^{-1}\langle)(\rangle \qquad \text{(I)}$$

$$\langle K \cup \bigcirc \rangle = (-a^2 - a^{-2})\langle K \rangle \qquad \text{(II)}$$

$$\langle \bigcirc \rangle = 1 \qquad \text{(III)}$$

Let us begin at the end. The third and simplest rule tells us that the Kauffman bracket for the diagram of the unknot $\bigcirc$ is equal to 1 (that is, the zero-order polynomial with the constant term 1). Rule II, in which K is any knot (or link), indicates how the value of $\langle K \rangle$ changes when one adds to it a trivial knot unlinked with K: this value is multiplied by a coefficient equal to $(-a^2 - a^{-2})$.

The first rule—which, despite its simplicity, is the fundamental rule underlying Kauffman's theory and this chapter—describes the connection between the brackets of the three links (or knots) symbolized by the three icons:

which differ only by a single small detail. More precisely, the icons denote three arbitrary links that are identical except for the segments of the strands indicated inside the dotted circles.

Those who have read the chapter devoted to the Conway surgical operations will doubtless have noticed the analogy that exists between Kauffman's rule I and skein relations. Let us simply recall what the icons for the skein relations look like:

How are they different from those in rule I? First, the knots considered by Kauffman are not oriented (there are no arrows). Consequently, there is only one type of crossing (two according to Conway), but two ways of smoothing (Conway's arrows impose a single smoothing, the same for the two different crossings). Having said that, rule I of Kauffman's theory is, like Conway's skein relations, nothing more than a very simple equation related to a local surgical operation.

Rules I–III make child's play of calculating a Kauffman's bracket from the diagram of a knot (or link): just apply rule I (taking care to note the intermediate equations obtained) until all the crossings are gone, then calculate, with the help of rules I and II, the bracket of the link of N disjoint circles–which is obviously equal to $(-a^2 - a^{-2})^{N-1}$.

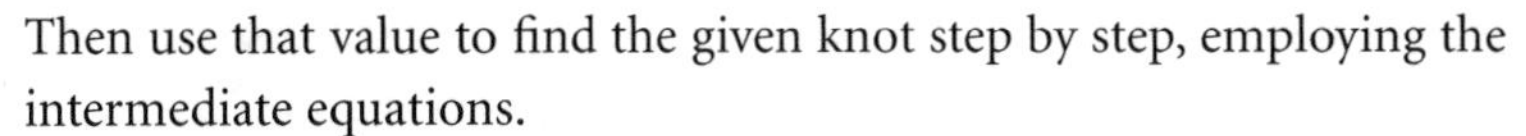

Then use that value to find the given knot step by step, employing the intermediate equations.

For example, for the unknot (using rule III) and the trivial two-component link (using rule II), we get:

$$\langle \bigcirc \rangle = 1$$
$$\langle \bigcirc\bigcirc \rangle = (-a^2 - a^{-2})$$

Following rule I and the preceding result,

$$\langle \infty \rangle = a\langle \bigcirc \rangle + a^{-1}\langle \bigcirc\bigcirc \rangle$$
$$= a \cdot 1 + (a^{-1})(-a^2 - a^{-2}) = -a^{-3}$$

In a similar way,

$$\langle \infty \rangle = -a^{-3}$$

For the Hopf link, using the preceding calculations, we get:

$$\langle \quad \rangle = a\langle \quad \rangle + a^{-1}\langle \quad \rangle$$
$$= a(-a^3) + a^{-1}(-a^{-3}) = -a^4 - a^{-4}$$

For trefoils, still using the preceding results, we obtain these formulas:

$$\langle \quad \rangle = a\langle \quad \rangle + a^{-1}\langle \quad \rangle$$
$$= a(a^6) + a^{-1}(-a^4 - a^{-4}) = a^7 - a^3 - a^{-5}$$

$$\langle \quad \rangle = a^{-7} - a^{-3} - a^5$$

Of course, the Kauffman bracket can only be used in knot theory if it is invariant, that is, if two diagrams of the same knot always have the same Kauffman bracket. This basic question is treated in detail in the following section, which is intended more specifically for those who have read the chapter on Reidemeister moves. The reader who has not (or who does not like mathematical proofs) will miss little by going directly to the subsequent paragraphs, where I finally introduce the Jones polynomial.

Invariance of the Kauffman Bracket

Thanks to Reidemeister's theorem, proving the invariance of the bracket requires only showing that its value does not change when the knot (or the link) undergoes Reidemeister moves. The reader will perhaps remember that there are three of those moves; take a look at Figure 3.1 in Chapter 3.

Let us begin with the second move, Ω_2. Using rule I several times and rule II once, we get:

$$\langle\ \rangle = a\langle\ \rangle + a^{-1}\langle\ \rangle$$

$$= a[a\langle\ \rangle + a^{-1}\langle\ \rangle] + a^{-1}[a\langle\ \rangle + a^{-1}\langle\ \rangle]$$

$$= [a^2 + a^{-2} + aa^{-1}(-a^2 - a^{-2})]\langle\ \rangle + aa^{-1}\langle\ \rangle$$

$$= \langle\ \rangle$$

Comparing the first and the last member of this series of equalities, we see that the invariance with respect to Ω_2 has been established. The attentive reader will have noticed the miraculous disappearance of the

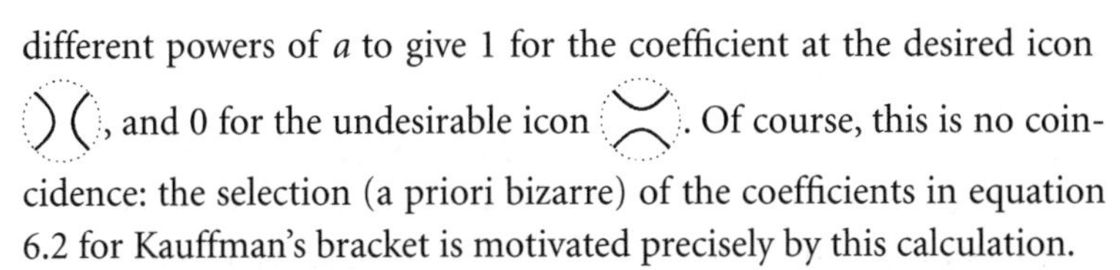

different powers of a to give 1 for the coefficient at the desired icon, and 0 for the undesirable icon. Of course, this is no coincidence: the selection (a priori bizarre) of the coefficients in equation 6.2 for Kauffman's bracket is motivated precisely by this calculation.

Inspired by our little victory, let us move on to verifying the invariance relative to Ω_3, the most complicated of the Reidemeister moves. Still using the basic rule I, we can write:

(6.3)

Obviously,

since these two diagrams are isotopic in the plane. Now, twice applying the invariance relative to Ω_2 (which we have just demonstrated), we obtain:

Comparing the righthand sides of the two equalities in equation 6.3 shows that they are equal term by term. That is therefore also true for the lefthand sides. But it is precisely this equality that expresses the invariance of the bracket with respect to Ω_3!

A Little Personal Digression

God knows I do not like exclamation points. I generally prefer Anglo-Saxon understatement to the exalted declarations of the Slavic soul. Yet I had to restrain myself from putting two exclamation points instead of just one at the end of the previous section. Why? Lovers of mathematics will understand. For everyone else: the emotion a mathematician experiences when he encounters (or discovers) something similar is close to what the art lover feels when he first looks at Michelangelo's *Creation* in the Sistine Chapel. Or better yet (in the case of a discovery), the euphoria that the conductor must experience when all the musicians and the choir, in the same breath that he instills and controls, repeat the "Ode to Joy" at the end of the fourth movement of Beethoven's *Ninth.*

Invariance of the Bracket (Continued)

To prove the invariance of Kauffman's bracket, it remains only to verify its invariance relative to the first Reidemeister move, Ω_1, the simplest of the three. Using rules I and II, we obtain:

$$\langle \quad \rangle = a\langle \quad \rangle + a^{-1}\langle \quad \rangle = \lambda\langle \quad \rangle$$

where $\lambda = a(-a^2 - a^{-2}) + a^{-1} = -a^3$.

Disaster! This blasted coefficient a^3 has refused to disappear (for the other little loop, we get the coefficient a^{-3}). Failed again! The Kauffman bracket is not invariant relative to Ω_1 and thus is not an invariant of knot isotopy. For example:

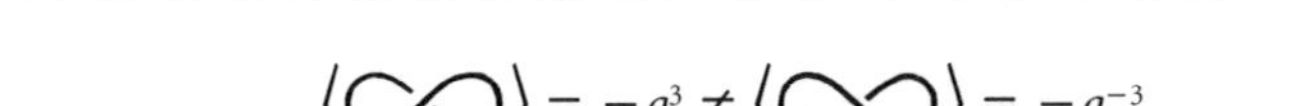

But the two diagrams both represent the trivial knot, and thus we *should* get:

$$\langle \infty \rangle = \langle \infty \rangle = \langle \bigcirc \rangle = 1$$

Should we plunge into despair?

Another Little Personal Digression

That's exactly what I did fifteen years ago when I was working on these same questions. The Jones polynomial and skein relations had made their appearance, and like many other mathematicians, I was playing with variants of these relations in the hope of finding invariants more sensitive than Jones's. Later, among my scribbles, I retrieved some formulas very close to Kauffman's rule I, and I remembered having been stymied by this same move Ω_1 (unable to get rid of a stubborn coefficient that just would not disappear) and dropping it.

But Kauffman persevered. Budding young researchers will grasp the moral of the story, even if perseverance doesn't always pay off—I am not at all sure that by continuing I would have managed to discover the fantastic little trick that allowed Louis Kauffman to succeed.

Kauffman's Trick and Jones's Polynomial

The starting point is obvious: since the coefficient $a^{\pm 3}$ refuses to go away, we will have to add a supplementary factor to our bracket whose purpose would be to rid us of this tiresome $a^{\pm 3}$. But how?

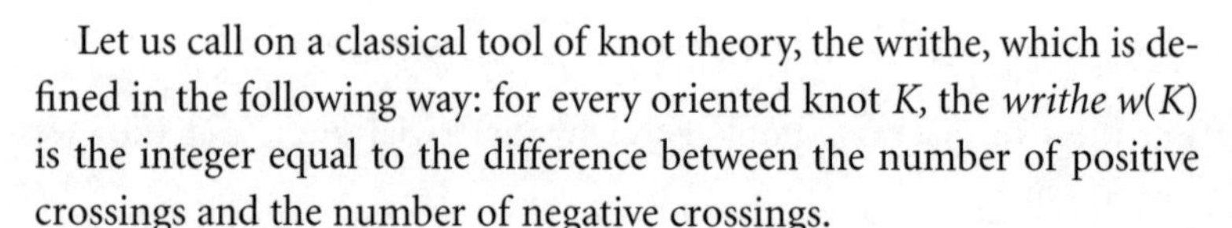

Let us call on a classical tool of knot theory, the writhe, which is defined in the following way: for every oriented knot K, the *writhe* $w(K)$ is the integer equal to the difference between the number of positive crossings and the number of negative crossings.

It is easy to see that the writhe is an invariant of the Reidemeister moves Ω_2 and Ω_3. In contrast, move Ω_1 changes the writhe: it adds 1 or -1 to it according to whether the eliminated loop is negative ( ) or positive ( ).

Still following Kauffman's lead, let us now define the Jones polynomial[4] of an oriented knot (or link) K by writing:

$$X(K) = (-a)^{-3w(K)}\langle\, |K|\, \rangle \tag{6.4}$$

where the nonoriented diagram $|K|$ is obtained from the oriented diagram K by forgetting its orientation (erasing the arrows) and where $\langle\cdot\rangle$ is the same Kauffman bracket that has caused us so much joy and sorrow.

Kauffman's trick is this factor $(-a)^{-3w(K)}$, which does a superb job of killing off the annoying coefficient $a^{\pm 3}$ resulting from move Ω_1. (I leave to the "mathematized" reader the pleasure of unraveling the details of this sophisticated murder, worthy of Agatha Christie.)

It is now obvious that the Jones polynomial is an isotopy invariant of knots (and links). Indeed, we have seen that the bracket $\langle\cdot\rangle$ as well as the factor $(-a)^{-3w(K)}$ are invariant with respect to Reidemeister moves Ω_2 and Ω_3, and everything goes smoothly with Ω_1 as well (lazy readers will have to take my word for it); the isotopy invariance of $X(\cdot)$ therefore follows from Reidemeister's theorem.

Before describing what Jones's polynomial contributes to knot theory, I am going to take advantage of the fact that the basic rule (I) of

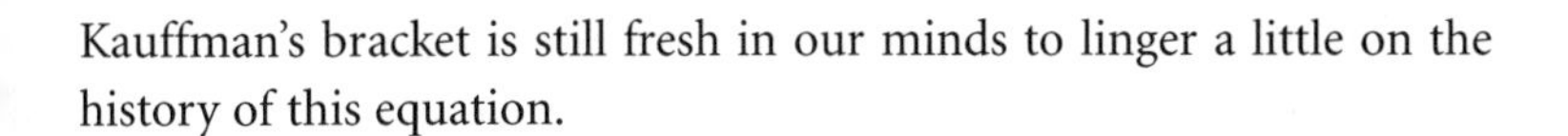

Kauffman's bracket is still fresh in our minds to linger a little on the history of this equation.

A New Digression—On Menhirs

No mathematician would deny Louis Kauffman the honor of having invented rule I, an insignificant-looking little formula whose fundamental character was immediately obvious. Yet, only a few years ago, Kauffman learned that he was not the first to come up with the formula: a specialist of ancient Celtic culture explained to him that sculptors who worked on menhirs six thousand years ago used exactly the same rule to alter the structures of connected ribbons (thus of knots and links) that decorated these burial stones. Readers will find motifs from rule I (can I still call it Kauffman's rule?) in the links of the ribbons carved on the menhir shown in the Preface (Figure P.5).

Rules for the Jones Polynomial

I have just demonstrated the first fundamental rule of Jones's polynomial:

(1) Two diagrams of the same knot (link) have the same Jones polynomial.

The second fundamental rule—whose proof derives from a fairly easy calculation based on the "Celtic rule" (I) of Kauffman's bracket and equation 6.4—is the skein relation for the Jones polynomial:

(2) $a^{-4}X(\;) - a^{4}X(\;) = (a^{2} - a^{-2})X(\;)$.

The two other rules are obtained directly from rules II and III for the bracket:

(3) $X(K \cup \bigcirc) = (-a - a^{-2})X(K)$.
(4) $X(\bigcirc) = 1$.

These rules are sufficient to calculate the Jones polynomial for specific knots and links. (In fact, one can show that rules (1)–(4) entirely determine the Jones polynomial.)

Let us do the calculation for the trefoil (to simplify the notations, I have written $a^{-4} = q$).

$$q^{-1}X(\text{trefoil}) - qX(\text{trivial knot}) = (q^{1/2} - q^{-1/2})X(\text{Hopf link})$$

We have obtained the trivial knot and the Hopf link. Let us do the calculation for the latter:

$$q^{-1}X(\text{two unlinked circles}) - qX(\text{Hopf link}) = (q^{1/2} - q^{-1/2})X(\text{circle})$$

Following rules (3) and (4),

$$X(\text{Hopf link}) = q^{-2}(q^{1/2} + q^{-1/2}) - q^{-1}(q^{1/2} - q^{-1/2}) = \\ = -q^{-1/2} - q^{-5/2}$$

Thus, for the trefoil:

$$X(\text{trefoil}) = q^{-2} + q^{-1}(q^{-1/2} + q^{-5/2})(q^{1/2} - q^{-1/2}) \\ = q^{-1} + q^{-3} - q^{-4}$$

Readers who have developed a taste for these calculations can check that the same result is obtained if we use equation 6.4 and the preceding calculation for the Kauffman bracket of the trefoil.

Similar calculations show that the knots in the little table of knots presented in the first chapter are all different. Do not think that proving this fact has no purpose other than to satisfy our mathematical pedantry. When Jones's polynomial appeared, its calculation for the knots with 13 crossings or fewer gave different values for all the knots, except for two specific knots with 11 crossings. This result was suspect, and a closer comparison of the diagrams of these knots (which look quite different) showed that they were in fact isotopic diagrams (diagrams of the same knot): the table was wrong.

This application (which caught the attention of specialists in knot theory) led Vaughan Jones to hope that his polynomial would be a complete invariant, at least for prime knots. Unfortunately, that wasn't to be: there are nonisotopic prime knots that have the same Jones polynomial (see, for example, Figure 5.8).

Nonetheless, for knot theory, the role of the Jones polynomial remains very important: it is a sensitive invariant—more sensitive, for example, than Alexander's polynomial. It distinguishes, among others, the right trefoil from the left trefoil, which Alexander's polynomial cannot do. Moreover, Jones himself and his followers found new versions of his polynomial that were even finer.

Having said that, I must point out that no one has succeeded in finding a complete invariant along these lines. Another effort, based on totally different ideas, is described in the next chapter.

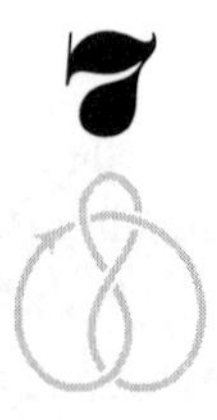

FINITE-ORDER INVARIANTS

(Vassiliev · 1990)

Victor Vassiliev should never have worked on knots. A student of Vladimir Arnold, and consequently a specialist in the theory of singularities (better known in the West under the media-friendly term *catastrophe theory*), he was unable to apply the techniques of this theory directly to knots—objects with a regular local structure, smooth and continuous, without the least hint of a catastrophe.

Perhaps a wise humanist whispered in his ear, "Since the singularity doesn't exist, it must be invented." Be that as it may, Vassiliev did invent it.

The concept is of disarming simplicity. Together with proper knots, Vassiliev explains, one must consider *singular knots;* these differ from true knots in that they possess double points, where one part of the knot cuts another part transversally.

In the planar representation of a knot, the appearance of double points barely differs from that of crossings, or . You might say that moving a true knot in space produces a "catastrophe"

when one part of the knot crosses another; at that instant, the knot becomes singular, then immediately reverts to an ordinary knot, but one that may be different from the initial knot. As an example, Figure 7.1 shows how the trefoil knot is changed (as it undergoes a catastrophe) into a singular knot with a single double point and then becomes a trivial knot.

Vassiliev jumbles together in one set (denoted by $\mathcal{F}$) ordinary knots with singular knots, which may possess any (finite) number of double points. Ordinary knots thus form a subset of $\mathcal{F}$ denoted by Σ_0, whereas the others form what is called the *discriminant* Σ. This is broken up into strata $\Sigma_1, \Sigma_2, \Sigma_3, \ldots$, constituted of singular knots with 1, 2, 3, . . . double points, respectively. It is in the vicinity of these strata that we are going to pursue our study of the invariants of knots.

Unfortunately, the stratified set $\mathcal{F} \supset \Sigma_0 \cup \Sigma_1 \cup \Sigma_2 \cup \ldots$ is of infinite dimension, so it is difficult to visualize. Nonetheless, I shall describe it very geometrically (though not very rigorously), shamelessly employing crude drawings, where the space (of infinite dimension!) $\mathcal{F}$ will be represented by a square: the one at the center of Figure 7.2. The points of $\mathcal{F}$ thus represent knots (singular or ordinary); around the square, we see more "realistic" representations of some of these knot-points, which show the process of deformation of a knot (the "figure eight knot") in Euclidean space $\mathbb{R}^3$. Inside the square, we see the path taken in the space $\mathcal{F}$ (a "symbolic" representation of the same deformation) by the mobile point $H \to G \to F \to D \to C \to B \to A \to \bigcirc$ corresponding to the knot undergoing deformation.

At the first catastrophic moment (when the double point 1 forms on the figure eight knot H), the mobile point cuts through the stratum Σ_1 (the stratum of the singular knots having exactly one double point)

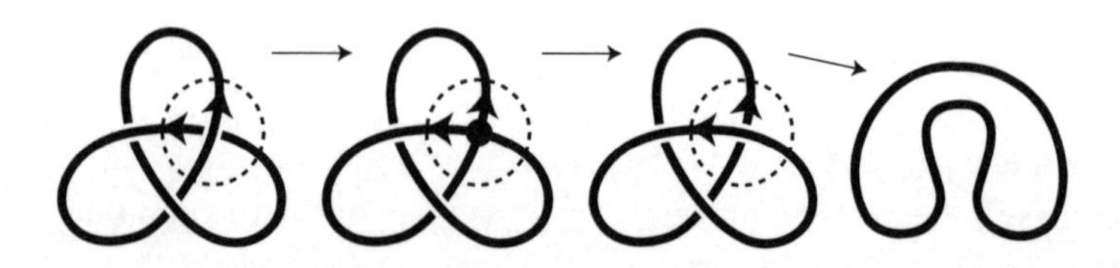

Figure 7.1. The trefoil becomes singular, then unknots.

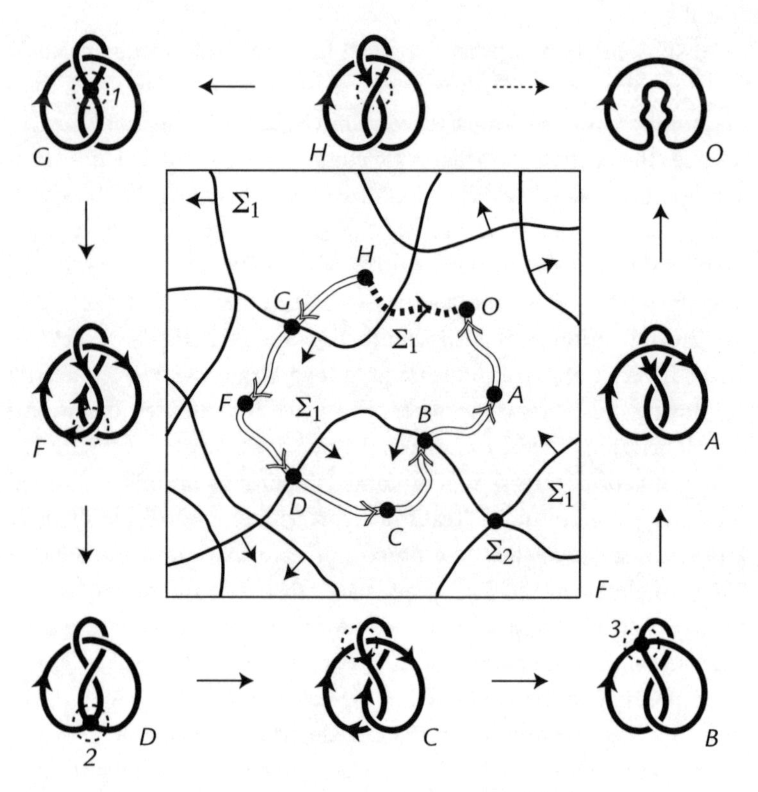

Figure 7.2. Deformation of a knot in space $\mathbb{R}^3$ and in $\mathscr{F}$.

at point G. The knot then becomes trivial (unknotted, and denoted F) and continues to deform until the second catastrophic instant, when a new double point, point 2, forms and disappears immediately, and the unknot changes into a trefoil. This event corresponds (in the symbolic representation inside the square) to a new crossing ($F \to D \to C$) through stratum Σ_1, but at another location (D). Finally, a new crossing through another part of the same stratum occurs ($C \to B \to A$), and the result is knot A, which is in fact the trivial knot (○).

Every Vassiliev invariant assigns to each knot (and in particular to singular knots) a certain numerical value. We will begin by giving a simple example of a specific Vassiliev invariant, which we will call v_0. To define it, we will assume it to be equal to zero for the trivial knot [$v_0(\bigcirc) = 0$] and postulate that each time the moving point M (which represents our knot) cuts across the stratum Σ_1 in the positive direction (that of the arrows[1] issuing from Σ_1), the value of $v_0(M)$ increases by 1. Thus one can easily calculate the value $v_0(H)$ of the selected Vassiliev invariant for the figure eight knot. For that, begin at point ○ (corresponding to the unknot ○), assuming $v_0(\bigcirc) = 0$, and follow the curve shown in Figure 7.2—$\bigcirc \to A \to B \to C \to D \to F \to G \to H$—by cutting across the stratum Σ_1 three times, once in the positive direction and twice in the negative direction, to obtain:

$$v_0(H) = 0 + 1 - 1 - 1 = -1$$

Right away, a question arises. Is the invariant in question correctly defined? Doesn't its value depend on the choice of the path that connects points ○ and H? Will one get the same result if one takes, for example, the path shown in a dotted line in the figure? Happily, the answer is

yes, both in this particular case [$v_0(H) = 0 - 1 = -1$] and in the general case. But this fundamental fact is not at all obvious, and it took all of Vassiliev's cleverness (and very sophisticated techniques from algebraic topology) to prove it.

What do these calculations tell us? First, that the figure eight knot cannot be unraveled, since its invariant is different from that of the unknot ($-1 \neq 0$). On the other hand, we have seen in passing that the trefoil is not trivial either [since $v_0(C) = 1 \neq 0$], and also that the figure eight knot is not equivalent to the trefoil [$v_0(H) = -1 \neq 1 = v_0(C)$].[2]

So this Vassiliev invariant does a good job of its basic task: it succeeds in distinguishing certain knots. It is, however, true that it is not a complete invariant: it cannot tell all knots apart; for example, simple calculations show that the value of v_0 for right and left trefoils is the same: the invariant cannot tell the difference between a trefoil and its mirror image. But this is not the only Vassiliev invariant. There are infinitely many of them! In particular, it is not all that difficult to find another Vassiliev invariant that can distinguish two trefoils; for that, we must move a little more deeply into the strata; in this case, descending to the neighborhood of stratum Σ_2.

But before proceeding from examples to general theory, I want to take a little rest from mathematical reasoning by making a short digression about the method used here.

Digression: Mathematical Sociology

Vassiliev's approach to knots could be called sociological. Instead of considering knots individually (as Vaughan Jones does, for example), he considers the space of all knots (singular or not) in which knots are only points and therefore have lost their intrinsic properties. Moreover, Vassiliev does not go looking for just one invariant—he wants to find all of them, to define the entire space of invariants. In the same way that classical sociology makes an abstraction of the personality of the people it studies, focusing only on their position in the social, economic, political, or other stratification, here the mathematical sociologist focuses only on the position of the point with respect to the stratification in the space $\mathcal{F}$: $\mathcal{F} \supset \Sigma_0 \cup \Sigma_1 \cup \Sigma_2 \cup \ldots$

This sociological approach in mathematics is not Vassiliev's invention. In the theory of singularities, it is due to René Thom, and it remains the weapon of choice of Vladimir Arnold and his school. Much earlier it was used by David Hilbert to create functional analysis (functions lose their own personality and become points in certain linear spaces), and by Samuel Eilenberg, Saunders MacLane, Alexander Grothendieck, and others, in a more striking manner, to lay the basis of *category theory.* This theory was ironically called "abstract nonsense" by more classically inclined mathematicians, perhaps to exorcise it: in the beginning, it seemed ready to devour mathematics whole. (Fortunately, it is clear today that nothing like that actually happened.)

But let us come back to Vassiliev and his singular knots. In this specific situation, the sociological approach—which we will soon delve into in detail—seems especially fruitful. All the information that is needed to define the invariants of knots can be found by exploring the strata $\Sigma_1, \Sigma_2, \ldots$ Like Vassiliev, we are going to try to find all of these invariants, by always going deeper and deeper, that is, by studying the strata Σ_n with greater and greater indices n. Since that requires a certain ease with mathematical reasoning, readers who do not have it can skip directly to the conclusion of the chapter.

A Brief Description of the General Theory

To recap: a singular knot K is any smooth curve[3] in space $\mathbb{R}^3$ whose only singularities are double points (in finite number), points where a part of the curve transversally cuts another part. Note that singular knots, like ordinary knots (see Figure 7.3), are oriented (marked with arrows).

For singular knots, as for ordinary knots, there is a natural relation of equivalence called *ambient isotopy:* two knots (possibly singular), K_1 and K_2, are isotopic if there is a homeomorphism of $\mathbb{R}^3$ (preserving the orientation) that sends K_1 to K_2, respecting the arrows (and the cyclic order of the branches with double points). (Then the terms *knot* or *singular knot* may stand both for a specific object and a class of isotopic equivalence—the reader can decide depending on the context.) We will denote by Σ_0 the set of (ordinary) knots and by Σ_n the set of singular knots with n double points.

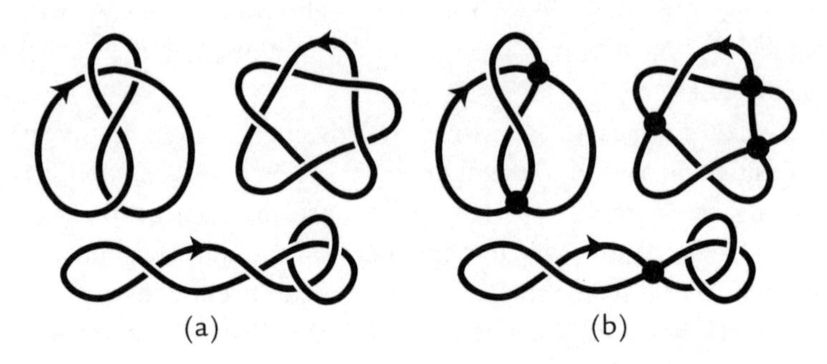

Figure 7.3. Ordinary knots (a) and singular knots (b).

Slightly displacing one of the branches of a singular knot near a double point smoothes the singularity with two different crossings:

$$\leftarrow \qquad \rightarrow$$

Recall that the smoothing to the left is called positive, the other negative.[4]

We say that a function $v\colon \mathscr{F} \to \mathbb{R}$ is a *Vassiliev invariant* (in the broad sense) if, for each double point in a singular knot, it satisfies the following relation:

$$v(\ ) = v(\ ) - v(\ ) \tag{7.1}$$

which means that the function v is applied to three knots identical everywhere but inside a little ball, where the knots appear exactly as shown in the three little dotted circles; the parts of the knot outside the ball are not shown explicitly, but they are the same for the three knots. The function v must be defined on the equivalence classes (elements of $\mathscr{F}$), and thus $v(K) = v(K')$ if K and K' belong to the same class.

From definition 7.1 we immediately deduce the so-called one-term relation:

$$v(\ ) = 0 \tag{7.2}$$

and the four-term relation:

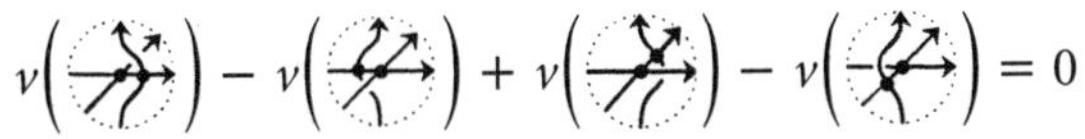

$$v(\ ) - v(\ ) + v(\ ) - v(\ ) = 0 \tag{7.3}$$

Actually, deducing equation 7.2 simply requires applying definition 7.1 once

$$v(\quad) = v(\quad) - v(\quad)$$

and noting that the two little loops obtained by solving the double point can be eliminated isotopically, giving two identical knots for which the difference between the invariants will indeed be zero. Providing relation 7.3 requires smoothing the four off-center double points, which gives eight terms (each with a single double point) that neatly cancel out two by two.

We say that a function $v\colon \mathcal{F} \to \mathbb{R}$ is a *Vassiliev invariant* of order less than or equal to n if it satisfies relation 7.1 and vanishes on all knots with $n + 1$ double points or more.[5]

The set V_n of all Vassiliev invariants of order less than or equal to n possesses an obvious vector space structure and has the inclusions $V_0 \subset V_1 \subset V_2 \subset V_3 \ldots$

> *Lemma* The value of the Vassiliev invariant of order less than or equal to n of a singular knot with exactly n double points does not vary when one (or several) crossings are changed to opposite crossings.

The idea behind the proof is very simple: according to equation 7.1, changing a crossing into an opposite crossing causes the value of the invariant v to make a jump equal to $v(\quad)$; but here this jump is zero, since the argument of v in this case is a singular knot with $n + 1$ double points.

An obvious consequence of the lemma is that zero-order invariants are all constants (in other words, $V_0 = \mathbb{R}$, the set of real numbers) and thus uninteresting. Indeed, we know that all knots can be unknotted by changing a certain number of crossings; and since these operations do not change the value of any zero-order invariant (according to the lemma), their value is equal to the value of the invariant of the unknot.

It can be shown almost as easily that there are no nonzero first-order invariants (in other words, $V_0 = V_1$), but, fortunately, the theory becomes nontrivial from the second order onward. To illustrate, we will distinguish among the elements of V_2 a specific invariant, denoted v_0, by taking it equal to 0 on the unknot and equal to 1 on the singular knot with the following two crossings: . Using the lemma, one can show that v_0 is well defined. The calculation, which uses definition 7.1 three times and the equality $v_0(\bigcirc) = 0$ three times, is shown in Figure 7.4.

In fact, what we have is the same invariant v_0 whose value for another knot, the figure eight knot, was calculated (without the validity of the calculation being rigorously established) at the beginning of this chapter. But this time our calculation is quite rigorous. Readers

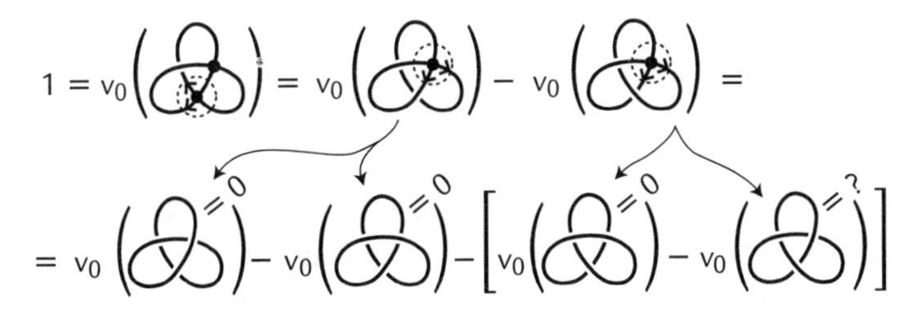

Figure 7.4. Calculating a second-order invariant of the trefoil.

who like these things can redo the calculation for the figure eight knot, as well as for other knots.

Gauss Diagrams and Kontsevich's Theorem

We are now going to divest ourselves of the geometry underlying our study of knot invariants and consider them in terms of a purely combinatory theory.

The lemma of the previous section tells us that the value of an nth-order invariant of a knot with n double points is unaffected by changes in the crossings. Thus, its value does not depend on the phenomenon of knotting; it depends only on the order (a combinatorial concept!) in which the double points appear when following the curve of the knot. We propose to code this order in the following way. Consider knot $K: S^1 \to \mathbb{R}^3$ with n double points. Proceeding around the circle S^1, we will label all the points sent to double points by the mapping of K, then join all the pairs of labeled points sent to the same double point by chords (Figure 7.5). The resulting configuration is called the *Gauss diagram* or the chord diagram of order n of the singular knot K.

Figure 7.6 shows all the Gauss diagrams of orders $n = 1, 2, 3$. Note that all the nonsingular knots have the same diagram (the circle without any cords). A good exercise for the reader who is hooked is to draw eight singular knots for which the eight diagrams of Figure 7.6 are the corresponding Gauss diagrams (there are of course many knots that correspond to the same diagram).

Now we will rewrite the one-term (equation 7.2) and the four-term (equation 7.3) relations in the language of Gauss diagrams. In this notation, a Gauss diagram actually stands for the value of an nth-order invariant (always the same one in the given formula) of one of the sin-

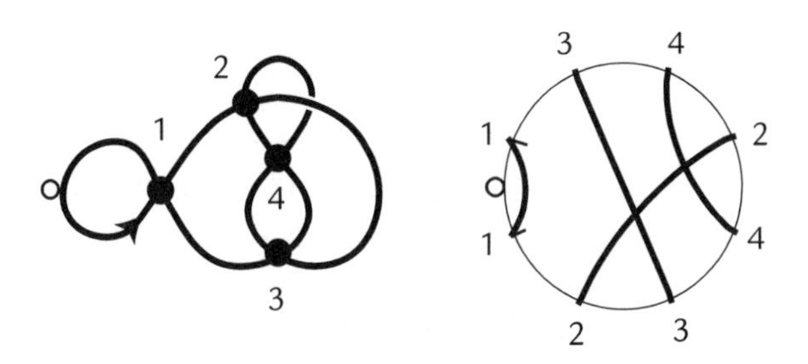

Figure 7.5. Gauss diagram of a singular knot.

$n = 1$: ; $n = 2$: , ;

$n = 3$: , , , , .

Figure 7.6. Gauss diagrams of order $n \leq 3$.

gular knots with n double points that corresponds to the diagram (which specific knot is chosen is immaterial because of the lemma). When there are several diagrams, we will not draw all their chords, but we will understand that the undrawn chords are identical for all the diagrams. In this way, we get:

= 0 − + − = 0 (7.4)

How are we to understand this notation? The first formula means that the value of each nth-order invariant for a singular knot with n double points that contains a little loop—with a double point (see equation 7.2)—is zero; thus, in this formula, I have omitted the terms $v(\ldots)$, and I have not drawn the other $n - 1$ chords of the diagram; moreover, it is understood that none of these chords can terminate on the little fat arc in the diagram. Similarly, the second formula describes the alternate sum of the values of the same nth-order invariant for four diagrams with n chords, but I have only drawn two in each diagram; the other $n - 2$ chords are exactly the same in all four. Moreover, it is understood that these supplementary chords cannot terminate on the little fat arc.

For example, for $n = 3$, the four-term relation specifically gives:

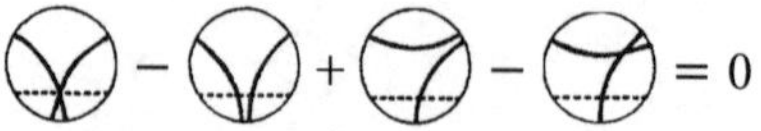

$$- \quad + \quad - \quad = 0 \qquad (7.5)$$

and since the third diagram vanishes—by virtue of the one-term relation (see the first equality in equation 7.4)—we get:

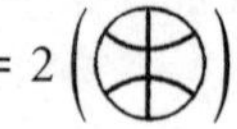

$$= 2\left(\quad \right) \qquad (7.6)$$

This relation can be envisaged as an equality in the vector space $\mathbb{R}$ of Gauss diagrams with three chords. More generally, one can consider the vector space $\mathcal{D}_n = \mathbb{R}(\Delta_n)$ of all finite linear combinations of Gauss diagrams $\mathcal{D} \in \Delta_n$; one can then write for $\mathcal{D}_n$ all the relations that follow from the one- and four-term relations and take the quotient of $\mathcal{D}_n$ by these relations. We obtain a vector space, which we denote by $\mathcal{A}_n$.

For example, for $n = 3$, one has dim $\mathcal{D}_3 = 5$ (Figure 7.6), but the one-term formula cancels the last three "basis vectors" of $\mathcal{D}_3$ shown in Figure 7.6; equation 7.6 expresses one of the two nonzero vectors remaining in terms of the other, so that dim$\mathcal{A}_3 = 1$. (The reader can check that dim $\mathcal{A}_4 = 3$.)

The main result of this combinatorial theory is that the space $\mathcal{A}_4$ completely describes nth-order Vassiliev invariants.

Kontsevich's theorem The vector space $V_n/V_n - 1$ of nth-order Vassiliev invariants is isomorphic to the space $\mathcal{A}_n$ of Gauss diagrams with n chords modulo the one-term and four-term relations.

The proof of this theorem, which is even more remarkable than the theorem itself, is unfortunately too long and difficult to be presented here (Bar-Natan, 1995). But from it we see that the study of the space of nth-order Vassiliev invariants (and the determination of their dimensions) can be reduced to a purely combinatorial calculation. True, this calculation is far from easy. But with the help of a supercomputer, Dror Bar-Natan at Harvard succeeded in finding the dimensions of the spaces $\mathcal{A}_n \cong V_n/V_{n-1}$ for $n = 0, 1, 2, \ldots, 9$. The values of these dimensions are 1, 0, 1, 1, 3, 4, 9, 14, 27, and 44, respectively.

The usefulness of this combinatorial theory (a more detailed study can be found in CDL, 1994) is not limited to calculating the dimensions of Vassiliev spaces; the theory can also be used to find the values of specific invariants of specific knots. For example, the invariant $v_3 \in V_3$, defined by the formulas $v_3\left(\bigcirc\right) = 0$ and $v_3\left(\circledast\right) = 1$, can be used to show that the right trefoil is not equivalent to its mirror image, the left trefoil. I'll leave this calculation to the deft reader.

Conclusion: Why Vassiliev Invariants?

Given Jones's polynomial invariants and those of his followers, is there really any need to invent more? Of course there is: all the polynomial invariants known to date are not complete, which means that two

nonequivalent knots can have the same polynomial invariant. In contrast, for Vassiliev invariants, the following assertion holds:

Conjecture Finite-order invariants classify knots; that is, for each pair of nonequivalent knots K_1 and K_2 there is a natural number $n \in \mathbb{N}$ and an invariant $v \in V_n$ such that $v(K_1) \neq v(K_2)$.

For the moment, this conjecture has neither been proved nor disproved.

Another justification for Vassiliev invariants is their universality: all the other invariants can be deduced from them. Thus, Joan Birman and Xiao-Song Lin, of Columbia University, demonstrated that the coefficients of Jones's and Kauffman's polynomials can be expressed in terms of Vassiliev invariants. Similarly, but at a more elementary level, I would suggest that readers of Chapter 5 prove, for example, that the coefficient at x^2 in Conway's polynomial $\nabla(N)$ for any knot N is a second-order Vassiliev invariant.

Today there are many other examples showing that Vassiliev's method makes it possible not only to obtain previously known invariants of knots, but also to define invariants—classical and new—for many other objects (and not only knots). But this aspect of the theory is beyond the scope of this book.

Finally—and this side of Vassiliev's approach seems the most interesting to me, for it is still developing—there are obvious and natural links (perhaps more than with the Jones and Kauffman polynomials) with physics. But I will talk more about that in the next and concluding chapter.

8

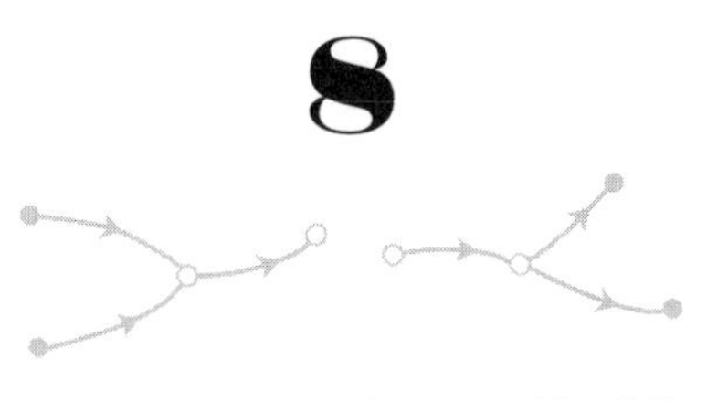

KNOTS AND PHYSICS

(Xxx? · 2004?)

This last chapter differs radically from the preceding ones. Their aim was to present the history of some of the (generally simple) basic ideas of knot theory and to describe the varied approaches to the central problem of the theory—that of classifying knots, most often tackled by means of different invariants. All the cases had to do with popularizing research results that have taken on a definitive shape. This final chapter, on the other hand, deals with research still ongoing, even research that is only barely sketched out.

Of course, one cannot make serious predictions about future scientific discoveries. But sometimes researchers working in a particular area have a premonition of an event. In everyday language, we say of such a situation (and usually after the fact), "The idea was in the air." The classical example—perhaps the most striking—is that of the independent discovery of non-Euclidean geometry by Janos Bolyai and Nicolai Lobachevski, which was anticipated by many others, and the unbelievable failure of Carl Gauss, who understood everything but did not dare.[1]

Is there "something in the air" today with regard to knots? It seems

to me that there is. I am not going to chance naming the area of mathematical physics where the event will occur, nor to name the future Lobachevski, nor predict (at least in any serious way) the date of the discovery: that is why the title of this chapter refers to Xxx with a question mark as the future discoverer and the fictitious date of 2004 (the end of the world, according to certain "specialists").

I will return briefly to predictions at the end of the chapter. But first, I wish to explain the sources of the remarkable symbiosis that already exists between knots and physics.

Coincidences

Connections linking knots, braids, statistical models, and quantum physics are based on a strange coincidence among five relationships that stem from totally distinct branches of knowledge:

- Artin's relation in braid groups (which I talked about in Chapter 2);
- one of the fundamental relations of the Hecke operator algebra;
- Reidemeister's third move (the focus of Chapter 3);
- the classical Yang-Baxter equation (one of the principal laws governing the evolution of what physicists call statistical models, which I talked about in Chapter 6);
- the Yang-Baxter quantum equation (which governs the behavior of elementary particles in certain situations).

These coincidences (visible to the naked eye without having to understand the relationships listed below in detail) are shown in part in Figure 8.1. At the left of the figure, we see the Yang-Baxter equation

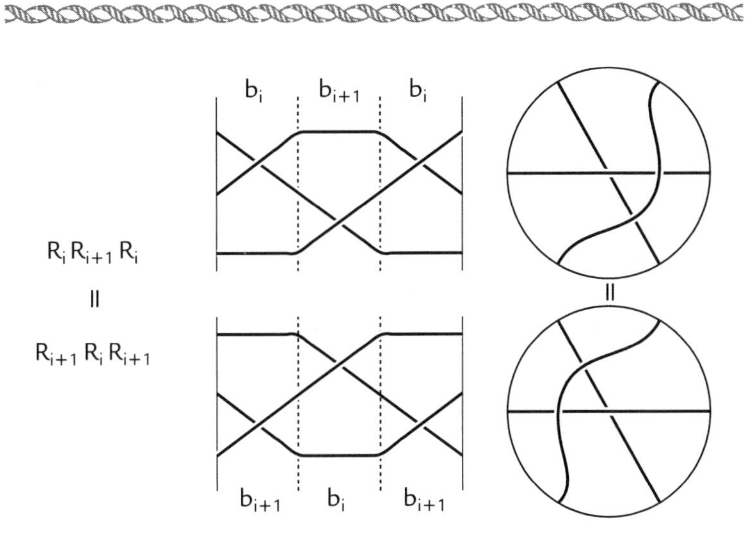

Figure 8.1. Three aspects of a single relationship.

$R_iR_{i+1}R_i = R_{i+1}R_iR_{i+1}$; at the center Artin's braid group, in algebraic form ($b_ib_{i+1}b_i = b_{i+1}b_ib_{i+1}$) and in graphical representation; and at the right, a drawing showing Reidemeister's third move. The two equations are in fact similar (just replace *b* by *R*, or vice versa), as are the two drawings (look closely!).

It was while exploiting these coincidences that the New Zealander Vaughan Jones; the Russians Vladimir Turaev, Nickolai Reshetikhin, Oleg Viro, and Vladimir Drinfeld; the Englishman W. B. Raymond Lickorish; the American Edward Witten; the Frenchman Pierre Vogel; and others discovered certain (profound? fortuitous?) connections between knot theory and several branches of physics.

A strange kind of statistical model devised by Louis Kauffman enabled him to describe the knot invariant actually discovered earlier by Jones—the famous Jones polynomial. Jones's original definition (not elaborated here) was based on braids and Hecke algebras (and thus on the coincidence between the Hecke and Artin relations). In Kauffman's approach (a version of which is presented in Chapter 6), it is the third Reidemeister move that plays the key role. Jones described a version of Potts's model (a statistical model very different from Kauffman's) based on the Yang-Baxter equation, which allowed him to rediscover his own polynomial in another way. Using certain solutions to the Yang-Baxter equation, Turaev discovered a whole series of polynomial knot invariants.

Should I go on? Wouldn't it be more reassuring to offer an explanation more specific and logical than "coincidence" for all these interdisciplinary links? Unfortunately, if a specific explanation exists, I do not know it. Yet there is indeed a more general explanation in the context of connections between mathematics and reality.

Digression: Coincidences and Mathematical Structure

All the sciences, natural or social, have an *object:* they purport to describe a certain part of reality, of real life. What is the object of mathematics?

The response is paradoxical: everything and nothing. "Nothing," because in mathematics one studies only abstractions, such as numbers, differential equations, polynomials, geometrical figures. Mathematicians have no specific object of study in material reality.[2] "Everything" because one can apply them to anything, any object that has *the same structure* as the abstraction in question. I am not going to try to

explain the meaning of the expression that appears above in italics, in the hope[3] that the reader will understand, for example (looking at Figure 8.1), that the Yang-Baxter equation has "the same structure" as the third Reidemeister move.

A (perhaps unexpected) consequence of this state of affairs is the importance of coincidences: if the structures of two objects happen "accidentally" to "coincide" (even if these objects have completely different origins), they are described by "the same mathematics," by the same theory. Thus, if the trace of an operator belonging to a Hecke algebra has the same properties as those of a knot invariant, why not produce a knot invariant by using this trace (which is what Jones did)? And if quantum particles, like knots, satisfy an equation that coincides with the Yang-Baxter equation, why not invent a theory of quantum particles using knot invariants (as did Sir Michael Atiyah, about whom I shall say more later)?

We have returned to concrete physical considerations related to knot theory; the general digression is therefore ended.

Statistical Models and Knot Polynomials

I talked about statistical models at the beginning of Chapter 6, in particular Ising's and Potts's models. Recall that they have to do with regular structures (for example, crystals) made up of atoms (with spins, say) that have simple local interactions (symbolized in the figures by the line segments joining the interacting atoms). Such a system X must have a partition function $Z(X)$ (which is the sum over all possible states of X of certain expressions that depend on the energies of local interactions); this function permits one to calculate the principal

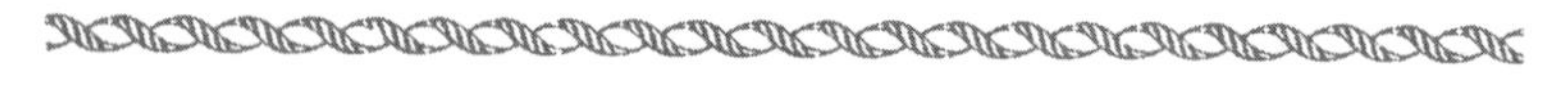

global parameters of the system (temperature, total energy) and to study phase transitions (such as passage from the liquid state to the solid state).

In Chapter 6, we saw how a kind of partition function allowed us to define and to calculate the Jones polynomial for knots. Actually, this function does not correspond to any real statistical model—rather, it is the fruit of the fertile imagination of Louis Kauffman. But it is most surprising that there exists a real statistical model, as noticed by Jones himself, with a genuine partition function, for directly constructing his polynomial. My immediate aim now is to describe this construction, without going into too much detail.

Given a planar diagram of a knot (or a link), begin by drawing its dual graph (or dual statistical model of the knot), as shown in Figure 8.2; you do that by alternately painting in black and white the parts of the plane delimited by the knot projection (taking care that the outside part be white), take the black regions for the vertices of the graph (or the atoms of the model) and join two vertices by a line or an edge (the interaction) if the black regions possess a common crossing. Moreover, declare the edges (interactions) to be positive or negative according to a convention that the reader can identify by looking at Figure 8.2.

Next, define the state of the system as an arbitrary function that assigns a spin to each atom, the spin taking only two values, which physicists call *up* and *down.* When the model is in a well-determined state $s \in S$ (S denotes the set of all possible states), the (local) interaction energy $E[s(v_1), s(v_2)]$ of the two atom-vertices joined by the edge $[v_1, v_2]$ is assumed to be equal to ± 1 if they have the same spin and to $a^{\pm 1}$ if the spins are opposite; choose the plus sign or the minus sign depend-

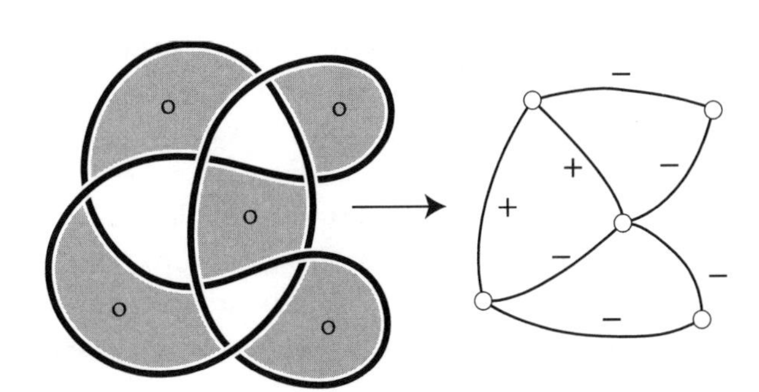

Figure 8.2. Dual graph of a knot.

ing on whether the edge (interaction) is positive or negative; here, a is the name of the variable of the polynomial (in a and a^{-1}) that you want to obtain. (It is this specific choice of the interaction energy of the atoms that is unique to Potts's model, the model of the phase transitions between water and ice.)

That done, we can define the partition function of the model by the formula:

$$Z(K) = \left(\frac{1}{\sqrt{2}}\right)\sum_{s \in S} \prod_{[v_i, v_j] \in A} E[s(v_i), s(v_j)]$$

where A is the set of all the edges.

Deriving Jones's polynomial from this partition function requires applying to it a variation of "Kauffman's trick" (described in detail in Chapter 6).[4]

Thus we see that Potts's model for freezing water brings us rather easily to the most famous invariant of knots. In analyzing this construction in surveys or popular articles, mathematicians have a tendency to enthuse over "the application of knot theory to statistical physics." Curious analysis! Knot theory has nothing to contribute to physics here—on the contrary, it is statistical physics that produces a construction that can be applied to mathematics. (To spare the self-esteem of mathematicians, recall that Jones's original construction—purely mathematical—preceded the "physical" construction just described.)

Of course, what is important here is not the rivalry between physicists and mathematicians, but this unexpected coincidence between two areas of knowledge, which are, a priori, very far apart. Let us move on to another coincidence, one that really deals with an application of knot theory to physics.

Kauffman's Bracket and Quantum Fields

I described Kauffman's bracket in Chapter 6, where it was used to define Jones's polynomial, the most famous invariant of knots. We are going to see that it can be used for something else entirely.

Recall that this bracket associates to every planar knot diagram K a polynomial $\langle K \rangle$ in a and a^{-1}, defined by a precise formula inspired by the partition functions of statistical models. It has already been noted that this equation (which we have no need of here) has no physical interpretation, at least in the framework of a realistic statistical model. It is in another branch of physics—topological quantum field theory—that it plays a role.

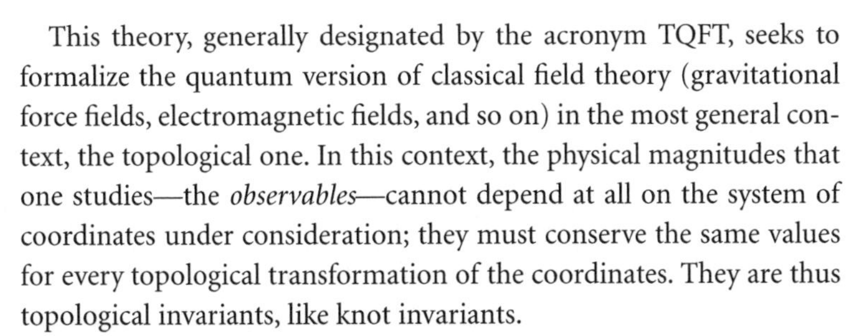

This theory, generally designated by the acronym TQFT, seeks to formalize the quantum version of classical field theory (gravitational force fields, electromagnetic fields, and so on) in the most general context, the topological one. In this context, the physical magnitudes that one studies—the *observables*—cannot depend at all on the system of coordinates under consideration; they must conserve the same values for every topological transformation of the coordinates. They are thus topological invariants, like knot invariants.

It was Witten's idea to use a generalization of the Jones polynomial (often called[5] the *Jones-Witten invariant*); he deserves the credit for finding it (and it won him the prestigious Fields medal) and for using it to construct a TQFT. This TQFT was simply a model whose dimension is 2 + 1, where 2 is the dimension of "space" and 1 that of "time," these three coordinates being mixed, as relativity demands. The model is thus a three-dimensional one that may contain knots, which in this context physicists call *Wilson lines.*

Later on, Michael Atiyah (also a Fields medal winner, but for earlier work) rethought Witten's model from a mathematical viewpoint and generalized it to create an axiomatic theory of TQFTs. Specifying this theory, Vogel and his coauthors constructed a whole series of examples of TQFTs, in which Kauffman's bracket actually plays the key role. There is no question here of explaining this theory and these examples—the mathematics required are too sophisticated. I will restrict myself to the context that features the bracket.

In this context, instead of a plane, think of a surface with a boundary, on which the diagram of a knot (or a link) is drawn. The strands of the knot may have ends at the boundary of the surface; a typical example is shown in Figure 8.3. Each of these diagrams is associated with a polynomial in a and a^{-1} that satisfies two very simple rules (already seen in Chapter 6):

$$\langle \times \rangle = a \langle \asymp \rangle + a^{-1} \langle)(\rangle$$

$$\langle K \cup \bigcirc \rangle = (-a^2 - a^{-2}) \langle K \rangle$$

Readers who recall this chapter will immediately recognize two fundamental properties of Kauffman's bracket. Note, to be done with coincidences, that a special case of this construction (when the surface is a disk) gives the so-called *Temperley-Lieb algebra,* an algebra of operators that satisfies the rules of Artin, Yang, Baxter, Reidemeister, Hecke, and so on.

I do not want to judge the interest of these TQFT models from the point of view of physical reality. Physicists take them very seriously,

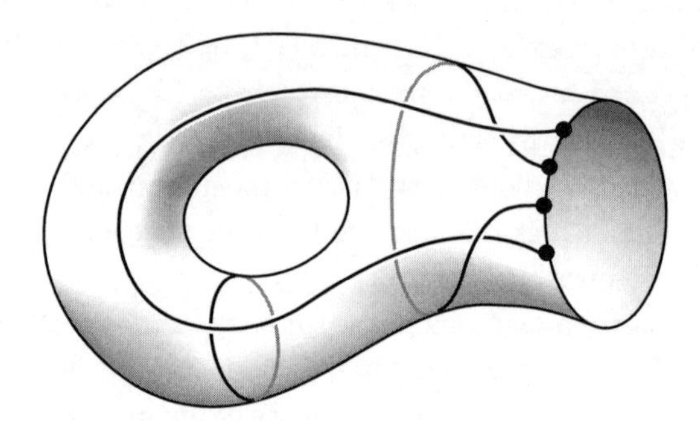

Figure 8.3. Diagram of links on a bounded surface.

but perhaps not as seriously as the (mathematical) idea of the quantum group, the study of which (in the context of its connection to knots) we will move on to next.

Quantum Groups as Machines for Making Invariants

Quantum groups appeared twenty years ago and today are an object of intense study by both mathematicians and physicists. Their formal definition, however, has little appeal: a quantum group is a set of abstract elements that must satisfy a whole list of formal algebraic axioms whose real meaning is not very obvious.

Rather than trying to explain quantum groups in detail, I will focus on their physical significance. First off, note that, despite their name, quantum groups are not groups at all; they are algebras, and even "bialgebras." That means that two operations are given for any set *Q*: a multiplication and a comultiplication. A multiplication, of course, associates to each pair of elements a well-defined element from *Q*—their product. The comultiplication does the opposite: it associates a pair of elements[6] from this set with a single element from *Q*—its coproduct. From the point of view of physics, these two operations correspond respectively to the fusion of two particles into a single one, and to the splitting of a single particle into two. I have tried to represent this correspondence graphically in Figure 8.4.

The operations (of multiplication and comultiplication) must satisfy some very obvious axioms (such as associativity) that endow *Q* with what mathematicians call a bialgebra structure.[7] These axioms are not all that restrictive, and there are so many quantum groups that one is led to consider a narrower class, for example, the class of

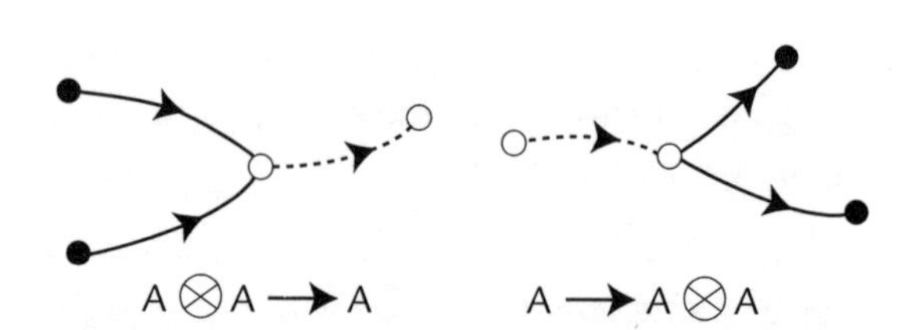

Figure 8.4. Product and coproduct of two particles.

quasi-triangular quantum groups defined by Drinfeld (another Fields medalist!). The axiom of quasi-triangularity implies that the Yang-Baxter equation holds for this class, and this, the reader will of course have guessed, provides the link between quasi-triangular quantum groups and knots. More precisely, the representations of these quantum groups allow us to define a lot of invariants, both new and familiar, one after the other. Quantum groups, as it were, are the truly scientific way to mass-produce knot invariants.

Vassiliev Invariants and Physics

As we saw in the preceding chapter, Vassiliev invariants are obtained by applying a very general construction, ideologically close to catastrophe theory, to knots. Can one give a fundamental physical meaning to the flip (the principal catastrophe, in the course of which the lower strand of a knot breaks through the upper strand, ending up on top)? No, it seems, at least not in any obvious way. The underlying physics is not evident in Vassiliev's work; it is buried in the algebraic and combinatorial structures characterizing invariants.

The thing is that the set V of invariants (which is in fact a vector

space) is not only endowed with a multiplication (obtained by multiplying the values of invariants, which are ordinary numbers), but also a comultiplication: $\Delta: V \to V \otimes V$. The latter is defined by means of the connected sum # of two knots via the following obvious formula:

$$(\Delta v)(K_1 \# K_2) = v(K_1) \cdot v(K_2)$$

It is easy to see that these two operations make a bialgebra out of V. Thus, right from the start, this "very physical" structure (fusing and splitting of particles) obviates the need to go looking "outside" for another algebraic object (such as the quasi-triangular quantum group for Jones-Witten-type invariants) to make "physical" invariants. This bialgebraic structure is inherent in Vassiliev invariants.

But there is more. First, at the analytical level, the Vassiliev invariants of a knot can be expressed via the admirable Kontsevich integral. In a certain sense, it is a generalization of the Gauss integral in electromagnetism, and thus should have a physical interpretation. What interpretation? Nobody knows.

Next, at the combinatorial level, the interpretation of Vassiliev invariants by chord diagrams (which I talked about in the previous chapter), another contribution from Maxim Kontsevich (yet another Fields medalist!), also lends algebra a physical orientation, several even. In particular, the algebra of Chinese characters (which until recently was called "the algebra of Feynman diagrams") is close to physical theory, as its (former) name indicates. But there, too, we are still at the stage of hope and speculation.

A final important point, also still not understood, is the four-term relation that I spoke about in Chapter 7. Dror Bar-Natan has exploited the fact that this formula is none other than a form of the classical

Jacobi identity to construct Vassiliev invariants using representations of Lie algebras. Will this coincidence between fundamental mathematical relations have a "physical" development?

Conclusion: Nothing Is Finished

At the beginning of this book, we saw how William Thomson's idea of using the knot to make a model of the atom almost a century and a half ago was the start of the theory of knots. Very recently, knot invariants, in particular Kauffman's bracket, became the basis for physics-oriented theories, such as topological quantum field theory. Where are we now?

Thomson's idea was ephemeral. From the viewpoint of actual physical reality, the significance of TQFTs (in the style of Witten, Atiyah, Vogal, Crane, Yetter) remains unclear, to say the least. Will interest in the connections between physics and knots be short-lived?

For specialists in knot theory, there is still a lot to do. For example, there is still no unknotting algorithm simple and efficient enough to be taught to a computer, and many other important questions have been shelved. For researchers in mathematical physics who look at knots from the side, many areas remain unexplored, particularly concerning Vassiliev invariants.

Finally, do not forget that aside from classical knots (three-dimensional curves in space), there are some less-studied "generalized knots," such as (two-dimensional) spheres in four-dimensional space (or surfaces, more generally speaking). According to Einstein, we live in four-dimensional space-time. According to specialists in string theory, the propagation of a particle can be modeled by sur-

faces. Is a quantum theory of gravitation hiding in there somewhere? Do Vassiliev invariants (which also must exist in this context) have a real physical interpretation?

Research always begins with a question, and hope. To conclude, I hope that the reader (and myself as well!) will again experience, in the context of knot theory, the incomparable joy of understanding inspired by a great discovery.

NOTES

Preface

1. For those who know, he is indeed the Vandermonde of the determinant.
2. Another strategy, simpler and perhaps no less efficient, is to thumb the book and choose which chapter to read according to the most interesting pictures.
3. The first definition of a polynomial invariant for knots, due to Alexander, was based on mathematical ideas that were very sophisticated for the time: homology theory and covering spaces.
4. In this case the physical interpretation uses the ideas of statistical physics.
5. Here I mean precise mathematical terms and not descriptions of TV horror films. Modern mathematical terminology, like that of theoretical physics, tends to favor everyday words over serious, scientific-sounding words.

1. Atoms and Knots

1. Differential geometry defines knots more neatly as "smooth closed curves," but this involves using the calculus.
2. Antoine is French for Antony.

2. Braided Knots

1. This terminology derives from what geometers call an inversion (in our case, a symmetry with respect to a small circle whose center is situated in

one of the regions bounded by the given Seifert circle). This inversion sends the center "to infinity" (and transforms the region into an infinite region).

2. This knot is numbered 5^2 in the tables of knots (Figure 1.6).
3. So called because they show up in all groups, not only in the group of braids B_n.
4. Among whom are certainly A. A. Markov, Joan Birman, and perhaps also William Thurston. And Emil Artin himself.

3. Planar Diagrams of Knots

1. Especially in "singularity theory" (often also called "catastrophe theory"), the basics of which were set down by Hassler Whitney between the First and Second World Wars and subsequently developed by René Thom, Vladimir Arnold, and their followers.
2. To the mathematically knowledgeable reader I should note that for a knot represented by a differentiable curve, proof of the analogous assertion is much more difficult and requires special techniques.
3. The reader may wonder how a computer can "see" knots. In fact, there are several effective ways of "coding" knots. For example, the one I use in my unknotting software gives the following description of the left trefoil: $1 + -2 - -3 + -1 - -2 + -3 - -1$. My computer understands it. Can the clever reader guess the principle behind the coding?
4. Perhaps I need to say here that for a projection of a given knot, the number of different applications of Ω_3 is finite. Moreover, be careful that the software doesn't make the mistake of following a specific application of Ω_3 by the inverse of the same application. Otherwise, the software may trigger a useless and infinite loop (in the computational sense of the term).
5. Actually, recent work by three American mathematicians (Joel Hass, Jeffrey Lagarias, Nicholas Pippinger) implies that the number of Reidemeister moves needed for unknotting is bounded, which means that in principle this algorithm either unknots a knot or declares that it cannot be unraveled

(since it hasn't succeeded in unraveling within the prescribed bound for the number of moves). Unfortunately, the bound is huge, and hence this argument is important only from the theoretical viewpoint. For practical purposes, however, a fast and usually effective unknotting software has recently been devised by the Russian I. Dynnikov.

4. The Arithmetic of Knots

1. Knots, remember, were defined as closed curves in three-dimensional space.
2. The pedantic reader will say that the box obtained is no longer cubic, and thus that the composite knot is not a true boxed knot. That is correct. By way of reward, let us leave to him or her the task of modifying the definition so that it will not be open to criticism; in particular, she/he should redefine the equivalence (isotopy) of knots and say that the knot, or the type of knot, is an equivalence class.
3. This idea will certainly occur to the reader who has assimilated the chapter on braids, for which this construction works perfectly well.
4. For the reader more familiar with mathematics, note that adopting the argument represented in Figure 4.5 specifically requires using another definition of knot equivalence, one that is based on the notion of homeomorphism.
5. It would be safer to say that we do not know whether an appropriate operation of addition exists for knots, since we know only that those who have searched for it have not managed to find one. We can, however, say with confidence that, if it exists, the geometric addition of knots is not simple—if it were, someone would have discovered it.

5. Surgery and Invariants

1. An angstrom (Å) is one ten-billionth of a meter.
2. This acronym is a flagrant injustice to two Polish mathematicians, Jozef

Przytycki and Pawel Traczyk, who made the same discovery at the same time but published later, not to mention many Russians who didn't publish because they considered the polynomial to be just a variant of the Jones polynomial. Another acronym, Lymphotu, was proposed later by Dror Bar-Natan. He gave credit to the Poles and to others (U = unknown discoverers), but the acronym didn't stick.

6. Jones's Polynomial and Spin Models

1. Or "intrinsic angular moment."
2. At issue here is a very theoretical model: "two-dimensional water." Of course, there is a more realistic three-dimensional model. We use "flat water" not only to simplify the drawing but also because it is the model most studied by physicists. But especially—as we will see later—we use it because the two-dimensional model relates to knots.
3. Generally, this polynomial can contain negative powers of *a*. Mathematicians call it a Laurent polynomial.
4. Strictly speaking, $X\langle\cdot\rangle$ is not the Jones polynomial; to obtain it requires changing the variable of the polynomial by writing $q = a^4$, but this is essentially only a change in notation.

7. Finite-Order Invariants

1. One chooses the direction of the arrows, called coorientation, in such a way that the following catastrophes (cutting across Σ_1)

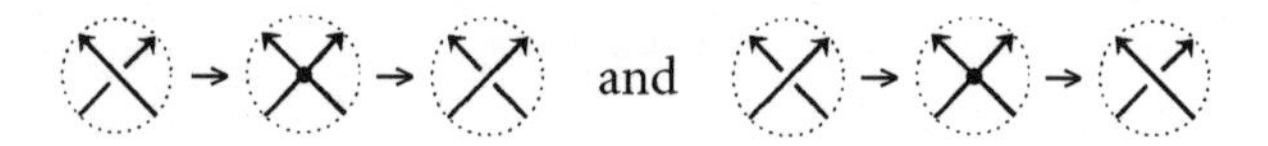

are positive and negative, respectively.
2. From a mathematical point of view, however, it must be emphasized that

we have not shown anything at all, and not only because we have not proved that v_0 is well defined. The problem is that our argumentation is based on the configuration of strata as it appears in Figure 7.2, but we know nothing of their actual configuration. Mathematized readers desiring a rigorous account of the preceding calculations will find one a little further on in this chapter. The others will have to take my word for it: the rigorous version of this calculation is really very simple, at least for anyone used to reasoning mathematically.

3. That is, every smooth mapping K of the circle S^1 into Euclidean space $\mathbb{R}^3$.
4. The positive (or negative, respectively) resolution is well defined: it is the one for which the traveller moving along the upper branch (following the arrow) sees the arrow of the lower branch pointing to the left (or right, respectively).
5. Vassiliev invariants are frequently also called *finite-order invariants* and sometimes *Gusarov-Vassiliev invariants,* because they were discovered independently by M. N. Gusarov from St. Petersburg (but he published his results only much later).

8. Knots and Physics

1. Gauss discovered the first principles of hyperbolic non-Euclidean geometry well in advance of Lobachevski and Bolyai but lacked the courage to publish this scandalous theory. (Lobachevski, who did publish it, quickly became the laughing stock of his contemporaries. And lack of recognition drove Bolyai to drink.) When Lobachevski's publication came out, Gauss had already created the differential geometry of surfaces, on which modelling hyperbolic geometry is mere child's play for a professional of Gauss's level. How could the brilliant Gauss, with all the tools at hand, have missed the discovery, as though he had suddenly been struck blind?
2. Barring the conviction—from a strictly Platonic point of view—that ab-

stractions, which belong to the world of ideas, are more real than the material world.

3. A famous attempt was made by Nicolas Bourbaki; I have nothing to say about it.
4. Actually, it is easy to show that the polynomial $Z(K)$ is invariant relative to the second and third Reidemeister moves, whereas the first move gives a superfluous factor that can be got rid of owing to another factor dependent on the writhe, exactly as shown in Chapter 6.
5. Unfairly. Witten's construction lacked mathematical rigor. It was N. Reshetikhin and V. Turaev who succeeded (thanks to another approach) in providing a correct mathematical definition for these invariants. That is why I prefer the expression Jones-Reshetikhin-Turaev-Witten invariants, despite its excessive length.
6. More precisely, a linear combination of pairs of elements.
7. Or a Hopf algebra structure.

WORKS CITED

Adams, C. 1994. *The Knot Book.* New York: Freeman.

Ashley, C. W. 1944. *The Ashley Book of Knots.* New York: Doubleday.

Bar-Natan, D. 1995. "On the Vassiliev Knot Invariants." *Topology,* 34:423–472.

CDL [Chmutov, S. V., S. V. Duzhin, and S. K. Lando.]. 1994. "Vassiliev Knot Invariants, I, II, III." In *Advances in Soviet Mathematics,* vol. 21, *Singularities and Curves,* ed. V. I. Arnold, pp. 117–126.

Dehornoy, P. 1997. "L'art de tresser." *Pour la science.* Special issue, pp. 68–74.

Haken, W. 1961. "Theorie der Normalflächen." *Acta Mathematica,* 105:245–375.

Jaworski, J., and I. Stewart. 1976. *Get Knotted.* London: Pan Books.

Jensen, D. 1966. "The Hagfish." *Scientific American,* 214:82–90.

Mercat, C. 1996. "Théorie des nœuds et enluminures celtes." *L'Ouvert,* no. 84.

Prasolov, V., and A. Sossinsky. 1997. *Knots, Links, Braids, and 3-Manifolds.* Providence: American Mathematical Society.

Rouse Ball, W. W. 1971. *Fun with String Figures.* New York: Dover.

Stewart, I. 1989. "Le polynôme de Jones." *Pour la science,* 146:94.

Thomson, W. 1867. "Hydrodynamics." *Proceedings of the Royal Society of Edinburgh,* 6:94–105.

Walker, J. 1985. "Cat's Cradles and Other Topologies Armed with a Two-Meter Loop of Flexible String." *Scientific American,* 252:138–143.

Wang, J. C. 1994. "Appendix. I: An Introduction to DNA Supercoiling and DNA Topoisomerase-catalyzed Linking Number Changes of Supercoiled DNA." *Advances in Pharmacology,* 29B:257–270.